CHARIOT IN INDIAN HISTORY

Chariot in Indian History

U.P. THAPLIYAL

ICHR

MANOHAR
2026

Reprinted 2026

First published 2022

ISBN 978-93-94262-12-6

Published by
Ajay Kumar Jain *for*
Manohar Publishers & Distributors
4753/23 Ansari Road, Daryaganj
New Delhi 110 002

Typeset by
Kohli Print
Delhi 110 051

Printed at
Replika Press Pvt. Ltd.

To the Sweet
and
Ever Cherished Memory
of
Our Beloved Son
Sheel
Who Left us Devastated
on
25 July 2012

Contents

List of Figures*

* Photographs collected from Archaeological Survey of India.

Foreword

Dr. U.P. Thapliyal has completed this important and fascinating study as part of the Indian Council of Historical Research 'Senior Academic Fellowship' scheme. In fact, the publication is a part of the Council's monograph series which was started many years ago. The chariot, in its various forms has been interwoven in our history. This monograph, through its eight chapters enriches the understanding of the history of chariots, their different purposes, including military and means of transportation with specific reference to Indian history from ancient to early medieval times. The first chapter presents the 'chariot' on the worldwide stage followed in the subsequent chapters in an engrossing manner to the chariot and its use in the course of Indian history. Based to a great extent on important sources such as ancient texts and scriptures that unfold in a chronological form, the result is a very readable and informative volume.

The study has 8 chapters beginning from the Vedic times. Not only has the author drawn the readers' attention to the chariot as a medium of warfare but to several other uses including the chariot's importance in temple life for example. The study has 18 key illustrations that put in perspective and explain the chariot and its utilization at length.

Dr. Thapliyal's hard work and sound historical method has resulted in a major contribution to a field hitherto understated, perhaps unresearched. The ICHR stands well rewarded with regard to having supported the study as part of its Senior Fellowship programme. My earnest congratulations to Dr. Thapliyal for a very good volume.

I must also compliment Prof. Kumar Ratnam, Member Secretary, Dr. Rajesh Kumar, Director (Journal, Publication & Library), and Dr. Md Naushad Ali, Deputy Director (Publication), for their efforts in the process of this publication. I must also thank Shri Ramesh Jain of Manohar Publishers for all his help and attention, resulting in a very good volume.

New Delhi
15 March 2022

RAGHUVENDRA TANWAR
Chairman, ICHR

Preface and Acknowledgements

While pursuing research on warfare in ancient India, I was perplexed by the preponderance of chariots in the Indian battlefield through the ages. It is for the first time in the Vedic literature that the evidence on chariot, as a vehicle of war and also as a means of transport, is met. By virtue of its importance in the life of the people chariot was, then, adored almost like a divinity.

But the moot point is, how the chariot was introduced in India. In the absence of any definitive evidence in the proto-historic India, as represented by the Indus Valley Civilization, it seems that it emerged in India some time after the eclipse of this civilization around 1500 BC.

The appearance of chariot in the Vedic age in a fully evolved form suggests that it was brought to India by immigrants who were well conversant with its use. Perhaps, a people with roots in Western Asia and Central Asia, were instrumental in this. It is notable that a branch of the Āryans called Mitannians, who knew all about chariots, had settled in Western Asia some time after 2000 BC. They spoke a language which was similar to Sanskrit and worshipped Vedic gods like Indra, Mitra, Varuṇa and Nāsatyas. The Vedic Āryans and Mitannians were in all probability the same people, who over the centuries settled in Syria, Iran and India.

For some time, I have been pondering over the improbability of Āryan chariot traversing through a treacherous mountainous track separating India and Afghanistan. It is likely that, they negotiated this inhospitable land mass after loading the dismantled chariot parts on the back of their horses. Alternately, a chariot may not have been brought in physically and the Āryans, who knew the art, manufactured them on arrival into India, as stated by Staal. But a more plausible explantion would be that, the Āryans brought in some prototypes of chariot along with the horses and produced many more after settling in India. The growing resistance of the natives may have accelerated the demand for chariots among the Aryans.

It is generally believed that, the chariot lost its primacy on the Indian battlefield after a poor showing in the battle of Hydaspes (Jhelum). But this is contradicted by the fact that Candragupta Maurya maintained a strong chariot corps and the *Arthaśāstra* of Kauṭilya considered the chariot as similarly strong and useful as the elephant in the battlefield.

There is also a viewpoint that the chariot faded out of the Indian battlefield soon after the foreign invaders like Indo-Greeks, Śakas and Kuṣāṇas settled in India because in their strategy of war, which involved sharp manoeuvres, there were no place for chariots. This also seems unlikely because some foreign rulers of India did maintain a chariot corps as a part of their fourfold army (*caturaṅgiṇī*). Further, there is substantial art and literary evidence to suggest that chariots remained an important arm during the early centuries of Christian era.

Though there is no numismatic or inscriptional evidence on the use of chariot during the Gupta age, the evidence furnished by the *Smṛtis*, *Purāṇas*, *Nīti* works and literary compositions of Kālidāsa, Bhāsa and Māgha is conclusive. Thus, the chariot survived in India as a war machine and also as a medium of transport till the Turkish invaders came and settled in the Indian subcontinent.

This dissertation has been broadly divided into eight chapters with a view to examine the subject comprehensively. The first chapter is introductory and describes the spread of chariot into various parts of the world, including India. The second chapter describes the pre-sence of chariots in the Vedic age. The third chapter deals with the role of the chariot in military rituals during post-Vedic period. The fourth chapter highlights the primacy of the chariot in the epic war. Though the epics were composed much after the *Vedas*, the references to chariots in them are quite in tune with the Vedic times. For this specific reason, we have placed the epics after the *Vedas*, chronologically. The fifth chapter deals with the *Mahājanapada* period lasting from sixth century BC to fourth century BC. Chapter six covers the Maurya-Śuṅga period dating between fourth and second century BC, and the subsequent Śaka-Kuṣāṇa phase. Chapter seven covers the period from fourth century AD to eighth century AD. The eighth chapter describes the eclipse of the chariot.

I avail of this opportunity to express my gratitude to my teachers Prof. Romila Thapar of Jawaharlal University, Dr. Devahuti of University of Delhi and Prof. J.S. Negi of Allahabad University – who taught me the lessons of history. I also bow my head to the sacred memory of my parents, Pt. Suresha Nand Thapliyal and Smt. Sarasvati Devi, who taught me the value of education. I also express my gratitude to my late brothers, Dwarika Prasad, Sharda Prasad and Bhagwati Prasad, who gave me the vision to see the world of letters. My debt to all of them in incalculable. My wife Sarojini also extended a helping hand.

My greateful thanks are due to the Archaeological Survey of India, for providing me useful photographs and also to the scholars mentioned in the bibliography for providing me vision to work on the dissertion 'Chariot in Indian History'.

I would also like to thank my friends Sarva Shri G.B. Singh, D.D. Nautiyal, H.N. Nautiyal, Bijai Singh Rawat, Umesh Chand Nailwal, Dr. Reva Dhanedhar, Karuna Srivastava and Dr. A.K. Mishra for helping me in the development of this book.

In the end I would like to record my grateful thanks to ICHR, who kindly granted me Senior Academic Fellowship to work on the project 'Chariot in Indian History'. I am also grateful to Manohar Publishers & Distributors for evincing interest in publishing this book.

New Delhi U.P. THAPLIYAL
16 December 2021

CHAPTER 1

Introduction: The Chariot on the World Stage

THE GENESIS

It is generally believed that chariots came into vogue in Mesopotamia around 3500-3000 BC (Figure 1.1). Called a battle-wagon, the Sumerians yoked wild asses to it as the horse had not been domesticated by then. Subseqently, they developed it into a military vehicle to transport the warrior and his weapons to the battlefield. This, in a way, revolutionized warfare in the ancient world by introducing the element of mobility.

Michael Stephenson has described the early usage of chariot as follows:

> The earliest chariot may have been used by Sumerians from Uruk in Mesopotamia around *c.*3500-3000 BCE. They were in their earliest incarnation, lumbering, solid-wheeled battle wagons, drawn by onagers, draft animals somewhere between horses and asses, and intended more as transport than shock weapons. Their primary function was to deliver into the heart of combat the high born warrior-duelist, who protected from arrows and sling-shots by an accompanying shield bearer, sought out his enemy peer.[1]

According to Yigal Yadin 'The invention and development of chariot was the most significant contribution to the art of warfare in the third millennium.[2]

The Sumerian war chariot was a heavy four-wheeled cart that carried a driver and a spearman. As a vehicle, it was not easy to manoeuvre because the solid wheels were originally fixed to the axles and the front axle could not swivel during turns. Because of its clumsiness and relative slowness of the draught animals that hauled it, the Sumerian battle-wagon was probably reserved for breaking up enemy formations in coordination with an infantry charge.[3]

Paul Kriwaezek, however, explains the earliest representation of chariot in the standard of Ur as follows:

> standard of Ur shows soldiers equipped for close quarter conflict. . . . In their support what are usually called chariots trundle up behind, though that word gives a quite false impression of their speed and manoeuvre-ability, given that these were clumsy, four wheeled, two-man vehicles drawn by asses. They cannot have moved very much faster than a man can walk. It may be better to think of them as mobile armouries, an interpretation supported by the large bucket at the front containing what looks like spare javelins.[4]

The origin of the evolved chariot is, however, associated with the domestication of the horse. To begin with, man hunted the horse for food as stated by Nic Fields:

> the horse was probably first domesticated on the steppes above the Black Sea, where semi-sedentary herdsmen used horses, like their cattle as a source of hide and food on the hoof. From the Srednij Stog culture of the Dnieper-Don, the bones of apparantly domesticated horses form the majority among those excavated from settlement sites dated to the fourth millennium BC. However, it remained an animal to be herded and eaten, rather than corralled and ridden.[5]

It was around 2000 BC that the horse was domesticated, and its potential as load-puller discovered. Indeed it turned out to be a far better substitute for bullock or the ass.

The wheel which had been invented by the Sumerians during the Uruk times (*c*.3100-2900 BC) was modified for use in the chariot.[6] These wheels were made out of three planks held together by wooden stouts and perhaps also by leather tyres secured by copper nails.[7] It is obvious that, these vehicles and their wheels could only be fashioned with the aid of saws and other metal tools.[8]

About the evolution of wheel Stephen Bertman's views are illuminating:

Sometime before 3000 BCE, the Sumerians became the first people in history to invent the wheel and dedicate it to the use of transportation. By the third millennium BCE, they were constructing small two and four-wheeled carts as well as covered wagons. The wheels of these vehicles were made of two half-discs of solid wood nailed together and covered with tires of leather like those of a cart found in the tomb of Puabi in Ur. Many centuries later, Assyrians manufactured metal tires out of sheets of copper, bronze and ultimately iron, which afforded the wooden wheels better protection. Originally, wheels of Sumerian vehicles were attached to the axle and turned with it. Later, they were designed to rotate separately around a regid axle to make cornering easier.

The discovery around 1500 BCE that wood could be bent with heat led to the development of wooden rims equipped with four to six spokes leading to a hub. Such lighter wheels meant faster vehicles.[9]

The introduction of spokes, ascribed to the proto-Indo-Iranians, ushered in a revolutionary change in wheel technology. The earliest known fully developed chariots (with spoked wheels) have been discovered from the burials of Andronovo (Timber Grave) sites of the Sintashta-Patrovka culture in Russia and Kazakhstan from around 2000 BC. The Andronovo culture over the next few centuries spread across the steppes from Urals and Tien Shan, corresponding to early Indo-Iranian cultures, which eventually spread to Iran and India in the course of second millennium BC. Piggott says that a light chariot with two spoked wheels appeared suddenly and almost simultaneously throughout a 'technological koine', which embraced all the land of civilization from Egypt to Mesopotamia.[10]

The material of which this body was built is unknown, but by analogy it is likely to have been of wicker work (as in Aegian and Celtic chariots), or perhaps leather (as in Egypt) on a light wooden framework.[11]

EARLY CHARIOT

The appearance of chariot could not have failed to be revolutionary, if only psychologically. Speed for human transport on land was suddenly multiplied by something like 10-from the two miles an hour for ox-transport to the 20 miles an hour reached with ease with a modern representation of an ancient Egyptian chariot with a pair of ponies, the chariot itself with its harness weighing only 75 lbs.[12]

It is said that, before being introduced in the battlefield, the chariot served the crucial need of the nomad pastoralists: to give a means of herding his flocks at a pace faster than his feet that could carry him and also to put him near – if not equal, terms of mobility with the predators, wolf and perhaps, bear and the large cat also, that harried their flanks.[13] The chariot warrior was, thus, born.

Perhaps, in the beginning the chariot was employed as a means of transport rather than a vehicle of war. But it helped the warrior to negotiate long distances and reach the battlefield untired. It also enabled him to carry heavy load of weapons. The warrior first brought the chariot to a halt, alighted and then joined the battle. Fighting while mounting the chariot could have become possible only after the institution of charioteer had come into existence.

In the words of Nic Fields,

> Early in the second millennium BC, the battlewagon underwent a revolutionary development that would turn this heavy compression-structure with solid disc wheels into an effective war machine. The development of steam-bent wood techniques not only allowed the construction of the spoked wheel with curved felloes, but also the manufacture of a light tension-structured chariot body. It would be this innovation that allowed the shifting of the single axle to the rear, which in turn made the chariot, as we can now properly call it, more manoeuvrable. Manoeuvrability was further enhanced by the use of the domesticated horse as a draught animal.[14]

In fact, with the introduction of the horse in the second millennium BC, the chariot became a swift-moving two-wheeled mobile firing platform that carried a driver, an archer with spare quivers, and a shield-bearer to protect the rest of the crew. Centuries later, the Persians added to the deadliness of the Babylonian and Assyrian chariot by attaching scythe-like blades to the rims of the wheels to chop up enemy troops during a charge.[15]

EARLY CHARIOTEERS

'Chariot and charioteers first appeared in the borderlands between the steppes and the civilized river lands', says John Keegan.[16] Elaborating further, he says that the language supplies some clue to this. The Hyksos who invaded Egypt originated on the semi-fertile northern fringe of the Arabian desert and spoke a Semitic language. The Hurrians and Kassites, who overthrew Hammurabi's Mesopotamian empire came from the mountainous headwaters of Tigris and Eupharates; the Kassites spoke an unidentified tongue classified as 'Asiatic', while the Hurrians and the Hittites, who established an empire in what today is Turkey, spoke Indo-European languages.

So too did the Āryan invaders of India, and it is possible that, the charioteers who founded the Shang dynasty in China may also have made their way thither from northern Iran, through perhaps, some proto-Iranian heartland in the Altai.[17] The *New Encyclopaedia Britannica* seems to support Keegan, while stating that, 'the chariot apparently originated in Mesopotamia in about 3000 BC; monuments from Ur and Tutub depict battle ponades that include heavy vehicles with solid wheels, their bodywork framed with wood and covered with skins.'[18] The development of this uncoordinate carriage into a useful military machine, however, took some 1,000 years. The introduction of horse as a draft animal, use of light spoked wheels, improvement in body structure of the chariot and horse accoutrement, etc., provided it with higher mobility and manoeuverability.

According to Keegan, 'the adoption of the war chariot and the imposition of the power of war charioteers throughout the centres of Eurasian civilization in the space of some 300 years (*c.*1700-1400 BC) is one of the most extraordinary episodes in world history'.[19] In the second millennium BC the chariot contributed to the victories of the Hyksos in Egypt, the Hittites in Anatolia, the Āryans in northern India and the Mycenaeans in Greece.[20] It is remarkable that, the light horse drawn chariot with spoked wheels heralded the beginning of a new epoch and appeared more or less simultaneously (around 1600-1500 BC) in Kassite Babylonia, Eighteenth Dynasty Egypt, the new kingdom of Mitanni, and in India. It made an appearance in Greece before 1500 BC, and in Crete about 1450 BC.[21]

MESOPOTAMIANS

The chariot, as an effective war machine, came into vogue around 1500 BC in different parts of the world. But the prototype of these chariots have to be traced to the battle carts of Early Dynastic kings (3000-2750 BC), towed by two or four onager hybrids (Figure 1.1).[22]

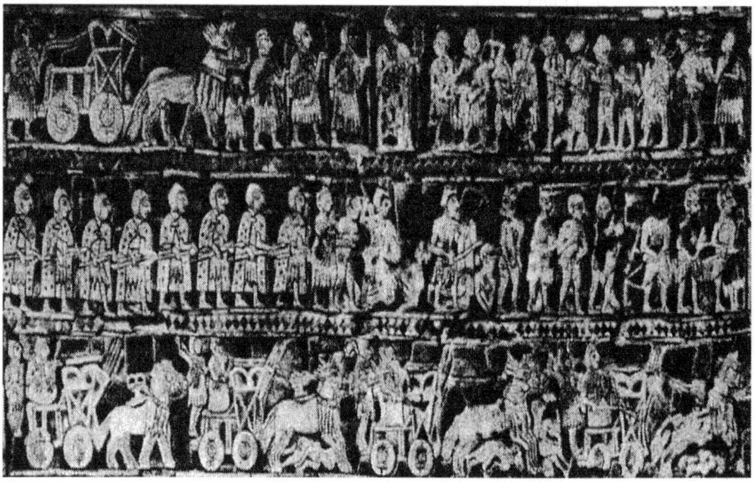

Figure 1.1: Early Chariot (Mesopotamia)

In the south, in the third millennium, to judge from the few representations of battle, the chariot were the elite arm. The four-wheeled carts of the standard of Ur were apparently towed by four animals and carried two fighters, who attacked with copper tipped javelins kept in a quiver at the front of the car. The ruler rides in one, and they are shown trampling down the prostrate enemy on cylinder seals. . . . Nevertheless the vehicles must have been unwieldly with their solid wooden wheels and primitive construction, and difficult to maintain.[23]

However, as the chariot spread across the world, it underwent change in make, shape and size (Figure 1.2).

Figure 1.2: Hittite Chariot – Qadesh relief

INDO-IRANIANS

It is believed that, the Āryans borrowed the idea of chariot from the Sumerians around 2000 BC. According to Stuart Piggott

> it looks, therefore, as if the battle-car was an invention of early dynastic Sumer and that its use was adopted, with other technological devices, such as metallurgy and the shaft-hole axe (and prabably the heptatonic scale in instrumental music), by the Indo-Europeans on the northernly fringes of the kingdoms of Sumer and Akkad soon after 2000 BC, given added speed and lightness by the use of horses and the invention of the spoked wheel, and spread by them in their expansion to east and west.[24]

They invented the spokes, did away with two of the four wheels and substituted horse for the ass.

This early Indo-Iranian chariot has been described in some detail by Stuart Piggott as follows:

> The Indo-Iranians were the first to develop the light, two-wheeled, spoked horse-drawn chariot. . . . The chariots were light and fragile weight, and therefore strength was sacrificed to speed; chariots were easily tipped over, splintered by arrows, and broken. On-the-spot repairs could be made with cord (but the cord was especially vulnerable to enemy arrows). The chariots axle was about four feet, four inches, the wheel about three feet

in diameter and the pole seven feet ten inches long. The box was made of leather on a light wood frame. Under it was a wooden axle fastened to the middle of the box by leather straps. The wheels were fitted to the projecting ends of the axle with lynch pins. The outer rim of the wheel was covered with metal. The wheel was spoked (probably six spokes). The pole angled up. The yoke was fixed to the pole with leather straps, and the yoke went over the backs of the horses and attached to their necks.[25]

MITTANIANS

A branch of these Āryans, called Mitannian, came to settle in Western Asia. They spoke a language similar to Sanskrit and worshipped the Indo-Āryan gods like Indra, Mitra, Varuṇa and Nāsatyas.[26] It is said that, 'this Āryan clan, moving westward brought with them their special knowledge of horse breeding, and that it was from them that the art was learnt by the people of Western Asia'.[27]

In this connection, it is notable that some tablets pertaining to fourteenth century BC mention Vedic numerals and terms that refer to the training of horses and the construction of chariots. A horse trainer Kikkuli, of Mitanni, whose language was extremely close to the Āryan language of Vedic Sanskrit, knew a great deal about taming, rearing, yoking and harnessing horses, their temperament and colours.[28] The treatise explains the essentials of training programme, which prepared the horse for drawing of chariots that were used for hunting or ceremonial purposes, but primarily for fighting battles. Similarities in chariot design in West Asia and India also tend to suggest a Western invasion of the subcontinent.[29]

The following observation by Asthur Cotterell also suggests a Western connection.

> In Mitanni, the Chief men were called *maryannu*, foreign experts in chariotry, upon whom the Mitannian kings relied. Although the language of Mitanni was Hurrian, a non-Indo-European tongue, there is no doubt of the presence of an Indo-European vocabulary in various Mitannian documents. The Hurrian plural, *maryannu*, meaning 'warriors' is *marya*, when singular and identical to Sanskrit *marya*, which means a 'young warrior'. Also these Mitanni works used to describe the colour of horses, for instance, parallel the languages of the *Ṛgveda*, India's oldest surviving text.[30]

It is likely that Sanskrit *maharathī* was the same as Hurrian *maryannu.*

Some scholars may be right in rejecting the invasion theory, but to my knowledge no one has advanced a convincing answer to Kaikkuli treatise and its Sanskrit affiliation in historical perspective. There are many more like John Keegan who strongly support the invasion theory.[31]

It may be mentioned here that, around this time the chariot was considered the foremost symbol of royalty and a prized possession by the rulers of Western Asia. This is revealed by a very interesting letter from Mitanni king Tushratta to Egyptian king Nimmureya. The salutation begins.

> For me all goes well, for you may all go well, for your household, for my sister, for the rest of your wives, for your sons, for your chariots, for your horses, for your warriors, for your country, and for whatever belongs to you, may all go very very well.[32]

HITTITES

The Hittites had mastered the art of chariotry by the late seventeenth century BC. They had also developed a chariot of lighter wheels, with four spokes rather than eight, and capable of carrying a crew of three, including a driver, a shield-bearer and an archer.

J.G. Macqueen has described the Hittite chariot as follows:

> The perfected chariot was a remarkably skilful piece of work, light in weight and extremely manoeuvrable at speed. The body consisted of a wooden frame covered with leather. This was mounted on a wide axle on which ran spoked wooden wheels. A pole ran forward from the underside of the body, on either side of which a horse was yoked. The superiority of Hittites in chariot-warfare lay not in their possession of this weapon (all their enemies had it too), but in their variation of the basic pattern to suit their own purposes.[33]

Each chariot carried a crew of three men clothed in leather garments that reached the mid-calf and a helmet. One man drove, the second wielded the offensive weapons (the lance and the bow) and the third provided protection with a shield.[34]

During their clash with Egyptians at Qadesh (Syria) in 1274 BC, Hittites deployed about 2,500 chariots. Detailed pictorial representations of this chariot battle at Qadesh have been left behind by Ramesses II (1279-1213 BC) on the walls of temples at Thebes, Luxor, etc. (Figure 1.3).[35]

Describing the Hittite conception of chariot warfare, Macqueen says that,

> To them a chariot formation was a heavy-weight assault force, which could sweep through and demolish infantry-lines in on organized charge. So we find that in Hittite chariots the principal weapon employed was the stabbing spear for action at close range. The axle was attached to the middle of the body rather than the rear. This meant that their vehicles were liable to overturn at speed, but the sacrifice in manoeuvrability was more than counterbalanced by the increase in firepower which resulted from it. For because of the forward mounting of the wheels, the Hittite chariot could carry a crew of three-a driver, a warrior and a soldier, who

Figure 1.3: Egyptian Chariot – Qadesh relief

during the charge held a shield to protect the other two. This extra weight was given to the charge and extra manpower was available in the hand to hand fighting which followed.[36]

Compared to Egyptian chariot, it was believed that, Hittite chariot was heavier since it had a central axle. But it was well suited to charge an enemy line, using the weight of the machine to crash through the opposing infantry. It was lethal, particularly in open terrain.[37]

HYKSOS

Hyksos, the West Asian warriors, who originated on the northern fringe of Arabian desert and spoke a Semitic language came to rule over northern Egypt between 1684 and 1555 BC. They set up a dynasty in Nile delta with capital at Avaris (modern-day Tell el-Dab'a). It is said that, Hyksos conquered Egypt with their chariot arm and lethal composite bow. Though Hyksos were finally defeated by the Egyptian king Ahmose and forced to migrate to Syria in 1555 BC, they left behind the legacy of chariot. Pharaohs used chariots to conquer many lands including Hyksos territory, subsequently.[38]

Hyksos maintained a well-organized chariot corps. The corps was divided into squadrons, each of which comprised twenty-five chariots and was commanded by a 'Charioteer of Residence'. Each chariot had two wheels, was drawn by two horses and carried two men, the driver and the warrior.[39]

It may be mentioned here, that chariot is widely claimed to be of Āryan origin since some technical terms associated with it are said to be of Indo-Āryan origin. To illustrate the *ratha* is the Indo-Iranian term for spoked wheel chariot of antiquity. It derives from a collective *ret-h-to*, a proto-Indo-European word *rot-o* for wheel, that also resulted in Latin *roto* and is also known from Germanic, Celtic and Baltic. The Sanskrit terms for wagon pole, harness, yoke and wheel-nave too have cognates with other branches of Indo-Europeans.[40]

The Āryan connection of the chariot in Egypt is confirmed by the fact that the Hyksos, who conquered Egypt with their chariot force, included a number of Hurrians, an Āryan people who hailed from Caspian region.[41]

EGYPTIANS

The chariot was introduced in Egypt in sixteenth century BC by Hyksos, the invaders from Asia Minor. Though Hyksos were soon driven out of Egypt, the chariot stayed on. The first direct reference to a horse drawn chariot in Egypt partains to Ahmose I (1549-1524 BC), the founder of Eighteenth Dynasty. The Pharaoh during an attack on Avaris, the capital of Hyksos, was carried by a chariot.[42] Tuthmosis I conquered many territories in fifteenth century BC using light riding war chariots.[43] Around 1500 BC the Egyptians invented the yoke saddle for their chariot-horses (Figure 1.4).

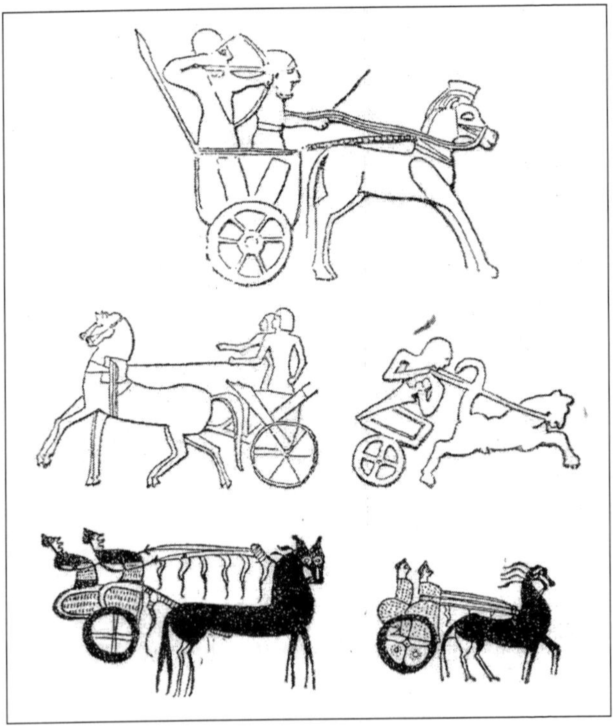

Figure 1.4: Some Early Chariots

The Egyptian kings maintained a strong chariot arm. The relief scene, depicting the battle of Qadesh at Luxor temple, depicts Rameses II alone in a chariot shooting arrows, unaccompanied by a charioteer or a shield-bearer. For protection, he is clad in a coat of mail while to dispense with the service of a charioteer he has the reins tied around his waist. His chariot is being pulled by one horse only. A normal Egyptian chariot, however, carried a crew of two persons a driver and a warrior, armed with bow and javelin.[44]

Here, it may be mentioned that an Egyptian chariot had its axle attached to the rear part of the body unlike the Hittite chariot which had the axle attached to the middle part. The Egyptian chariot, thus, provided for maximum manoeuvrability at high speed. Macqueen concludes that, 'the Egyptians regarded chariots as highly mobile firing platform from which long and medium range missiles could be dispatched in a manner which could cause the maximum of confusion in the enemy ranks'.[45] Among the Hittites, on the other hand, 'a chariot formation was a heavy weight assault force which could sweep through and demolish infantry-lines in an organized charge.'[46]

The chariot was used for various purposes by the Egyptians, including ceremonial rituals, hunting, sporting, transport and war. Those deployed in war were provided with spiked or bladed wheels, which could cut through the adversary.[47] However, among the Egyptians, chariots played a secondary role as these were mainly designed to protect their own infantry from the onslaught of enemy chariots. The chariots, richly decorated with gold, were perhaps meant for ceremonial use. Three such chariots have been discovered from the tomb of Tutankhamun.

CHINESE

The chariot was introduced in China during the rule of Xia dynasty in seventeenth century BC. But, 'it is possible that the charioteers who founded the Shang dynasty in China may have made their way thither from northern Iran also, perhaps from some proto-Iranian heartland in Altai'.[48] According to A. Cotterell, 'today a people canvassed for the introduction of the chariot from the Caucasus into north China are the Tocharians'.[49] The burial sites discovered in Henan province of China suggest that the scions of the royal family were buried along with the chariot, the horse and the charioteer, beside other things.

During the Shang times, the chariot was drawn by two horses, though occasionally four horses were yoked. The crew included a charioteer, an archer and a halbardier. The charioteer stood in the middle of the chariot-box, directly behind the draught pole; to his right was stationed the halbardier, to his left the chief chariot warrior who was armed with a composite bow. There were occasions when the bow-man and halbardier swapped sides for reasons of safety.[50]

Better harnessing seems always to have been typical of the Chinese chariot. Their invention of the breast-strap harness, the prototype of modern collar one, permitted an animal to throw its entire weight against a load because the harness pulled on the shoulders, leaving the throat unaffected.[51] Chariot, however, remained an important war machine in China up to third century BC.

The Chinese military thinker Sun Tzu, who lived in sixth century BC, refers to the wide use of chariots in war. The following extracts from his *The Art of War* substantiate this:

> (In deployment) with one thousand fast chariots, one thousand leather-covered chariots and one hundred thousand armored troops to be provisioned over one thousand *li*.[52]
>
> (In a chariot battle) when more than ten chariots are captured, reward him who first captures one. Then change their flags and pennants. When the chariots are mixed together, ride them, supply the captives and care for them.[53]

The Chinese rulers observed the custom of presenting war chariots to their army commanders. One such chariot is described as:

> Bronze fittings, with a decorated cover on the handrail, scarlet breast trapping of soft leather for the horses, a tiger skin canopy with a brown lining, painted leather axle coverings with gilded brake fittings, a fish skin quiver, bright harnessing and a scarlet pennant with two bronze bells.[54]

ASSYRIANS

The Assyrians had developed a fine military machine, with a powerful chariot component. An extended network of highways enabled their armies to campaign as far as 300 miles from their base. The horse elements of their army could march as fast as thirty miles in a day.[55] The Assyrian king Sennacherib fought a grim battle with the Elamites in 691 BC, and it has been recorded by a court scribe as follows:

My prancing steeds, trained to harness, plunged into their welling blood as into a river; the wheels of my battle chariots were bespattered with blood and filth. I filled the plain with corpses of the warriors like herbage. . . . (There were) chariots with their horses whose riders had been slain as they came into the fierce battle so that they were loose by themselves; those horses kept going back and forth all over (the battlefield). . . . As to the Sheikhs of the Chaldaeans (Elamite allies), panic from my onslaught overwhelmed them like a demon. They abandoned their tents and fled for their lives, crushing the corpses of their troops as they went. . . . (In their terror) they passed scalding urine and voided their excrement in their chariots.[56]

EARLY GREEKS

The chariot had reportedly made appearance in Greece before 1500 BC but little information is available about it earlier to Homeric age (850-800 BC). However, by thirteenth century BC these had become popular in the Greek world so as to play a significant role in an extended war between mainland Greece and Troy.[57] In *Iliad*, there is a graphic description of a war chariot:

Hera (queen of Zeus) harnessed the golden bridled team and Hebe (her daughter) quickly rolled the wheels to the chariot, paired wheels with their eight spokes all bronze, and bolted them on at both ends of the iron axle. Fine wheels with fellies of solid, deathless gold and round them running rims of bronze clamped fast-a marvel to behold! The silver hubs spin round on either side of the chariot's woven body, gold and silver lashings strapping it light, double rails sweeping along its deep full curves and the yoke-pole jutting forward, gleaming silver. There at the tip, she bound the gorgeous golden yoke, she fastened the gorgeous golden breast straps next and under the yoke queen Hera led the horses, racers blazing for war and the piercing shrieks of battle.[58]

Summing up the information on chariot gathered from the writings of Homer, Emile Mireaux says:

it was a light, rustic affair on which two men could stand upright. The main part of it was a body, of which the base consisted of a wooden frame a little over six feet wide – broader than it was long – and which was rounded in front. Over this frame were stretched solid leather thongs, which made a fairly elastic floor, good enough to deaden the shocks and jolts. A double barrier surrounded three sides of the car, to prevent one's falling forward or sideways over the wheels. . . . These chariots could cover a distance of about fifty miles from sunrise to sunset.[59]

Describing the make of a Greek chariot-wheel in detail Mireaux adds,

A fairly short pole was attached to the body and this was supported by an axle seven cubit long (over 9 ft). The wheels were mounted on this axle when it was ready to start. These wheels comprised a hub in which were fixed eight spokes supporting a rim; the rim consisted of four strips of wood cut in the direction of the grain, warped in the fire and bent in a circle, and encased one within the other – the whole wheel being consolidated by means of a ring of metal, which has been first expanded in the fire and which then, as it cooled clamped itself tightly round the rim. The whole carriage did not stand very high, as the wheels were only about three feet in diameter.[60]

CHARIOT RACE

Though chariot racing had come into vogue in Greece in the thirteenth century BC, it became a feature of funeral games during the time of Homer.[61] *Iliad* graphically describes the funeral games dedicated to Patroclus.[62] It seems that these funeral games matured into true sport subsequently.

In the plain below the stadium Olympia was built a special hippodroma for horse races. . . . The culminating events of the games were the chariot races, with two or four horses running abreast. Often ten four-horsed chariots competed together; and as each had to negotiate twenty-three turns around the posts at the end of the course, accidents were the chief thrills of the game; in one race with forty starters a single chariot finished.[63]

It is notable that, participation in a chariot race demanded skilful handling of teams particularly at the turning post. The *Iliad* (13.518) actually refers to this in detail by way of a dialogue between Nestor of Pylos and his son Antilochus:

> Drive close to it (turning post), and bear thy car and horses hard upon it, throwing weight upon the well-laced car. A little to their left; and then call on the off horse, voice and whip as well, and give him the rein, and let the near horses hug the post so tightly that the nave of thy stout wheel appears to graze it; but beware of hitting the stone, for fear of smashing up the car and injuring the horses.

In conclusion, it appears that the chariot first came into vogue as a means of transport around 3000 BC. To begin with, it was pulled by asses or onagers and carried men and material to the battlefield at a faster speed. The yoking of the horse turned it into a fast moving war machine and the emergence of charioteer made it far more effective in the battlefield. By 1500 BC, the chariot had matured into an effective vehicle of war in all cradles of civilization in Asia and Europe. The chariot emerged in India about the same time.

PROTO-HISTORIC INDIA

There is no positive evidence on the presence of horse and for that matter the chariot in the proto-historic India, as represented by the Indus Valley Civilization. In the absence of definitive evidence serious differences persist among the archaeologists. According to B.B. Lal, 'the much touted argument about the absence of horse from the Harappan Civilization has no validity in the light of new evidence regarding its presence'.[64] The evidence adduced by him to butteress his theory includes the following[65]:

1. Saudor Bookonye of the Archaeological Institute, Budapest, after examining the faunal remains concerned had declared as far back in 1993 that 'the domestic nature of the Surkoṭaḍā horse (a Haṛappan site in Kachchha) is undoubtful'.

2. A terracotta model of the horse found by E.J.H. Mackay in his excavations at Moheṅjo-Daṛo is displayed in the National Museum.
3. Lothal, a famous Haṛappan site is Gujarat, have yielded a terracotta figure of horse.
4. The faunal evidence from other Haṛappan sites like Kālībaṅgan, Rūpnagar, etc., is equally compelling.
5. Spoked wheels (of the chariots), widely referred to in the Vedic texts, have been discovered from the mature Haṛappan levels at sites like Banawālī, Kālībaṅgan, Dholāvīrā, Surkoṭaḍā, Rākhīgaṛhī and many other sites from Gujarat.

Disagreeing with Lal, many archaeologists hold that there is no evidence of fast transport like chariot in the Indus Civilization. D.P. Agrawal holds that

> the appearance of domesticated horse in the Old World was closely linked to the development of light weight chariots, which play a central role in the *Ṛgveda.* The oldest archaeological remains of chariots are from east and west of the Ural mountains, where they appear by *c.*2000 BC. Chariots like these were high tech creations.[66]

Mallory states that the horse was embedded in the culture of early Europeans and all the related languages used a term derived from proto-Indo-European *ekwas* for horse. This animal, it is certain, was a native of Eurasian steppes and nowhere else; the wild ancestor from which it was domesticated belonged to these steppes.[67]

Now the question arises as to when and how did the horse appear in India. Agrawal is perhaps right in stating that,

> the introduction of chariot and horse is one marker for the earliest possible date of the *Ṛgveda.* The ancient Iranian and Vedic word for chariot was coined sometime around 2000 BCE – about when chariot first appeared – but before these languages split into two. A good guess is that this occurred in the steppe belt of Russia and Kazakhstan, which is where we find the first remains of chariot.[68]

Bookonye also supports the extra Indian origin of horse while stating that, 'horse reached the Indian subcontinent in an already domesticated form coming from the Inner Asiatic horse domestication centres'.[69]

The absence of horse and chariot in the proto-historic India seems to be confirmed by the fact that

> not one clear example of horse bones exists in Indus excavations or elsewhere in north India before *c.*2000 BCE. All contrary claims arise from evidence from ditches, erosional deposits, pits or horse graves originating hundred or even thousand years later than Haṛappan Civilization. Remains of horse claimed to be early Haṛappan by archaeologists in the 1930s were not documented well enough to let us distinguish between horses, hemiones or asses.[70]

The discovery of a solid wheel of a toy chariot in albaster carved with geometric and other designs and three terracotta horses at Lothal[71] contradicts the widely held view that Indus Valley people were unfamiliar with the horse and chariot. However, there is no evidence to suggest that they used a wheeled vehicle like the chariot in war. S.D. Singh who has dealt with the issue at some length concludes, that there is no proof that the people of the Indus or Lothal ever made use of the wheeled vehicle as an instrument of war.[72] In this connection, the following quote from Piggott is also instructive:

> The Haṛappan civilization while fully cognizant of wheeled vehicles, does not seems to have made use of them in warfare-indeed as we have seen evidence for any military organization for defence or offence within the Haṛappan empire, is strangely lacking and the Āryan chariots clearly owe nothing to native Indian traditions. They appear as startling innovations, and to trace their ancestry, we must look to the west.[73]

He further adds that it is proved 'by a correlation of the Vedic texts with the evidence of archaeology that the war chariot of the Āryans in India was essentially the same vehicle as that known from Homeric Greece, or in Celtic Britain'.[74]

M. Sparreboom in his study, *Chariots in the Veda* also rejects Haṛappan association,

> we are at least on firm ground if we state that the Indo-Āryan tribes, when invading Iran and north-western India were chariot experts. . . . So most

likely the Vedic tribes did not derive their skills in chariotry from the Haṛappans.[75]

Cotterell, asserting that the Indo-Āryans introduced the chariot in India argues that

> the civilization of the Indus Valley was familiar with wheeled transport, model carts having been excavated at Moheṅjo-Daṛo, Haṛappā and Chanhu-Daṛo. A wagon from Chanhu-Daṛo with four solid wheels recalls the original construction method used in Sumer and Elam, an ancient state in present day Iran known for its chariotry. The spoked wheels of later Elamite chariots had the luxury of bronze tyres. Except perhaps at Lothal, an ancient port in modern Gujarat, nothing suggests the use of a wheeled vehicle as an instrument of war before the arrival of Indo-Āryans. But the model horses found at Lothal could have drawn a carriage instead of a chariot. The lack of evidence for the war chariot in the Indus Civilization is totally unlike the position of the Indo-Āryans, who were obsessed with charioteering.[76]

On the other side rejecting the thesis that chariot was introduced in India from outside Frawley says:

> Horses have been found, as well as wheeled toys that suggest the use of carts and chariots as much as in the Vedas. In the *Ṛgveda* itself the enemies of the Vedic people are also stated to have horses. The enemies of Sudās, after their defeat, offer him their horses heads as a tribute (*Ṛv*. 7.18.19). Horses are in fact one of the objects that the Āryans win from the Dasyus (*Ṛv*. 3.34.9). Hence there is nothing in the Vedas making the horse or the chariot a special possession of the Vedic people.[77]

But the knowledge of horse does not necessarily imply that the pre-Vedic Indians also knew the chariot.

SOME ANCIENT BATTLES ASSOCIATED WITH CHARIOTS

Battle of Megiddo

The first recorded chariot battle in the history of mankind took place at Megiddo, a fortified city midway between the Mediterranean and Jordan Valley, in northern Canaan (modern-day Tel Megiddo, Israel) in 1469 BC. Here a major encounter took place between the Egyptian king Tuthmosis III and the Canaanites under Hyksos leadership. The Egyptian army comprising around 20,000 men and 1,000 chariots used surprise and mobility to win a swift victory. Canaanite chariotry could not hold the Egyptian attack and took shelter behind the walls of Megiddo.

After a siege of seven months, Megiddo surrendered. All rebel leaders were made *prisoners* and forced to swear allegiance to the Egyptian ruler. The booty captured after the war included:

> 340 living prisoners and 83 hands; 2041 horses, 191 foals, 6 stallions and . . . colts; a golden chariot belonging to enemy; a fine chariot worked with gold belonging to the ruler of Megiddo . . . and 892 chariots of his wretched army – in total 924 and many more things.[78]

Battle of Qadesh

A second major chariot battle was fought between the Egyptian king Rameses II and Hittite ruler Muwatalli II at Qadesh on the banks of the Orontes River in Syria in 1274 BC. Some 5,000 chariots participated in this battle. An account of this battle is recorded on temple walls at Karnak and other religious sites. Rameses made a concerted charge, broke into Hittite formations and forced them to plunge into the river abandoning their chariots.[79]

The Hittites and Egyptians altogether, adopted different tactics in the chariot war, the former depended mainly on spear as a weapon of aggression, while the latter shot arrows from a distance in a defensive way.[80]

Battle of Qargar

The battle took place in 853 BC between the Assyrian king Neo-Shalmancsar III and a confederation of twelve kings (Neo-Assyrian and Israel). The confederates fielded an army of 3,940 chariots, 1,900 horsemen, 52,400 foot soldiers and 1,000 camels. The Neo-Assyrian king launched the attack to subdue the lands west of the Euphrates with an army totalling to about 60,000. No doubt, his army might have included a comparable chariot component of some 4,000 vehicles. Perhaps, the battle of Qargar saw the greatest assemblage of chariots in the history of war. The battle is said to have ended in a draw.[81]

Battle of Chengpu

The battle of Chengpu in China was fought between the armies of the states of Jin and Chu in 632 BC. The Jin forces were led by Wen and the Chu forces by Tzu Yee. In a carefully planned manoeuvre, in which the dust raised by a squadron of Jin chariots, pulling the trees, led the Chu commander to believe that it was a confused withdrawal. Caught in the trap, Tzu Yee was defeated by an elite force of chariot warriors under Wen.[82]

After the victory, Wen presented the reigning king with all the booty, including 400 four-horse teams of chariots and 1,000 foot soldiers. The king honoured him with the presentations of a large ceremonial chariot, a war chariot, bows, arrows, etc.[83]

Battle of Bi

The battle of Bi was again fought between the Jin and Chu states in 595 BC. The cause of the battle was the subjugation of Chen by Chu state, which was greatly resented by Jin state. Though the Chen soon made peace with Chu, the Jin could not digest the affront and assumed an aggressive posture.

The Chu Army took to the field while personal guard of thirty chariots stationed on the right and left wings protected the commander Duke Chuan. On learning that Jin were advancing, the Chu army rushed forward, its chariots racing along with its foot soldiers swooping down on the Jin army. The Jin army retreated in panic to the river to cross over, leading to many casualties.[84]

Battle of An

The battle of An was fought in 589 BC between the combined forces of the states of Jin, Wey and Lu and the ruler of Qi. In this battle, the Jin army was led by three commanders, Hsi Kao being the senior most.

The battle was joined by the Qi ruler in undue haste, even without waiting for his horses to be protected with armour. On the other side, Hsi Kao, the Jin commander, let his chariots race forward, despite severe wounds. Qi army was routed and chased away. Defeated in war, the Qi ruler offered surrender.[85]

Battle of Gaugamela

The battle of Gaugamela was fought between the Persian king Darius III and Alexander in 331 BC, near the modern Iraqi city of Erbil. In this battle, the Persian king deployed scythed chariots provided with razor-sharp blades on their axles and on the front side of the yoke pole. The objective was to breakthrough the Macedonian infantry lines by rushing through them. Regular chariots were deployed on the left wing for an outflanking move. It is said that while preparing for battle, Darius III levelled the battle-ground and also built runways for his chariots. However, Alexander successfully encountered these Persian moves, led a cavalry charge against Darius and forced him to flee in his chariot.[86] This was perhaps the last important chariot battle fought in West Asia in the ancient history.

NOTES

1. M. Stephenson, *The Last Full Measure,* pp. 15-16.
2. Y. Yadin, *The Art of Warfare in Biblical Lands,* p. 36.
3. S. Bertman, *Handbook to Life in Ancient Mesopotamia,* p. 266.
4. P. Kriwaezek, *Babylon: Mesopotamia and the Birth of Civilization,* pp. 91-2.
5. Nic Fields, *Bronze Age War Chariots,* p. 6. According to Piggott, the natural habitat of wild horse and its early domestication was on the South Russian steppe. *Wagon, Chariot and Carriage,* p. 48.
6. G. Childe, *New Light on the Most Ancient East,* p. 130.
7. Ibid., p.149; According to Nic Fields 'to give them greater strength, and to allow for a greater diameter, the wheels were made not of a single wooden plank but of three sections. These were bound together with external slots or ropes, dowelled together internally, and the whole consolidated with either a leather, reed or wood tyre secured by copper hobnails. These not only strengthened the wheel but also gave greater stability to the vehicle by acting as studs that gripped the ground.' op. cit. p. 11.
8. Ibid., p. 151
9. S. Bertman, op. cit., p. 254; Also see A. Cotterell, *Chariot,* p. 41.
10. S. Piggott, *Prehistoric India,* p. 278.

11. Ibid.
12. Ibid., p. 159.
13. J. Keegan, *A History of Warfare*, p. 165.
14. Nic Fields, op. cit., p. 12.
15. S. Bertman, op. cit., p. 266.
16. J. Keegan, op. cit., p. 159.
17. Ibid., pp. 167-8.
18. *The New Encyclopaedia Britannica*, vol. 3, p. 102.
19. J. Keegan, op. cit., p. 155.
20. *The New Encyclopaedia Britannica*, op. cit.
21. S. Piggott, op. cit., p. 276.
22. J.N. Postgate, *Early Mesopotamia*, pp. 146, 166.
23. Ibid., pp. 246, 248; This widespread of chariot culture is best described as follows:

    The Indo-European tribes had trained teams of horses, developed a light weight chariot, and combined the chariot with their expertise in the use of composite bow to produce a highly mobile platform from which they could deliver accurate, rapid fire. Their massed chariots overwhelmed native people from western Europe to the borders of China. One tribe (from the group of Indo-Europeans known as Indo-Iranians) invaded northern Mesopatomia, joined one of the native people there, the Hurrians, and formed the Kingdom of Mitanni; another tribe invaded southern Mesopotamia and Elam, where they formed a part of the new Kassite kingdom; and a third tribe divided into Āryans and the Iranians: the Iranians occupied the Iranian plateau, the Āryans invaded India and destroyed the thousand year old Harappan civilization. Another tribe, the Argives, invaded and conquered Greece. (A. Kaegi, *The Life in Ancient East*, p. 13.)
24. S. Piggott, op. cit., p. 278.
25. Ibid.
26. Paul Thieme, 'The "Āryan" Gods of the Mitanni Treaties', *JAOS*, vol. 80, 1960, p. 301: 'The discovery of Āryan looking names of [Mitanni] princes on cuneiform documents in Akkadian from the second half of the second millennium BC (chiefly tablets from Boġazkoy and El-Amarna), several doubtlessly Āryan words in Kaikkuli treatise in Hattite on horse training (numerals: *aika-* "one", *tera-* "three", panza- "five", *satta-* "seven", *naua-* "nine"; appelatives: *uarttana*-circuit, course [in which horses move when being trained], *asua-* "horse" and finally a series of names of Āryan divinities on a Mitanni-Hatti and a Hatti-Mitanni treaty poses a number of prob-

lems.' Ibid. There is also a view that names of Mitanni Āryan kings like Saustattar and Parsatattar denote chariot warriors and their Sanskrit form could be *Savyasthātar* and *Prasthātar*. Parpola, A., 'The Nāsatyas, The Chariot and Proto-Āryan Religion', paper read at 50th International Conference of Eastern Studies at Kyoto, 2005, p. 19.

27. Gurney, *The Hittites*, p. 105.
28. Frits Staal, *Discovering the Vedas*, p. 11; All Mitanni transliterations are very close to Vedic words, See Chart at p. 12.
29. A. Cotterell, *Chariot*, p. 2.
30. Ibid., p. 67.
31. J. Keegan, op. cit., p. 168.
32. Moran qoted by A. Parpola, op. cit., p. 13.
33. J.G. Macqueen, *The Hittites*, pp. 57-8.
34. Alfred S. Bradford, *With Arrow, Sword and Spear*, p. 30.
35. Nic Fields, op. cit., p. 47.
36. J.G. Macqueen, op. cit., p. 58.
37. Nic Fields, op. cit., p. 47.
38. Barbara Mertz, *Temples, Tombs and Hieroglyphs*, p. 136.
39. R. David, *Handbbok to Life in Ancient Egypt*, p. 229.
40. K.K. Thapliyal, *Wikipedia.*
41. R. David, op. cit., p. 239; Also see J.G. Macqueen, op. cit., p. 57.
42. Nic Fields, op. cit., p. 7.
43. Winifred Holmes, *She was Queen of Egypt*, p. 10.
44. R. Mathews, and G. Roemer, *Ancient Perspectives on Egypt*, p. 95; also pp. 85, 94.
45. J.G. Macqueen, op. cit., p. 58.
46. Ibid.
47. J. Keegan, op. cit., pp. 91-3.
48. Ibid., p. 168.
49. A. Cotterell, op. cit., p. 205.
50. Ibid., p. 195.
51. Ibid., p. 42.
52. Sun Tzu, *The Art of War*, Chap. 2; Evidently, Chinese army deployed two types of chariots in war – light and heavy. The light ones might have been used in attacking role while heavy ones in defending role. No other country seems to have followed this practice.
53. Ibid.
54. A. Cotterell, op. cit., p. 211.
55. Quoted by J. Keegan, op. cit., p. 172.
56. S. Piggott, op. cit. p. 274.

57. J. Keegan, op. cit., p. 240.
58. *Iliad*, 5.824-40.
59. E. Mireaux, *Daily Life in the Time of Homer*, pp. 72-3.
60. Ibid., p. 73.
61. Ibid., p. 236.
62. Ibid., p. 239.
63. Will Durant, *The Life of Greece*, p. 215.
64. B.B. Lal, *The Homeland of the Āryans*, p. 80.
65. Ibid., p. 81
66. D.P. Agrawal, *The Indus Civilization-An Inter-disciplinary Perspective*, p. 153.
67. J.P. Mallory, *In Search of Indo-European: Language, Archaeology and Myth*, p. 298.
68. D.P. Agrawal, op. cit., p. 153.
69. S. Bookonye, *South Asian Studies*, vol. 13, pp. 297-307.
70. Patel and Meadow, *Frontline*, vol. 17, issue no. 20, 2000.
71. A. Ghosh, *Indian Archaeology, 1959-60: A Review*, p. 18; Also Pls. 15B and 15E.
72. S.D. Singh, *Ancient Indian Warfare with Special Reference to the Vedic Period*, p. 24.
73. S. Piggott, *Prehistoric India*, p. 275.
74. Ibid., p. 273.
75. M. Sparreboom, *Chariots in the Veda*, p. 4.
76. A. Cotterell, op. cit., p. 168.
77. David Frawley, *Gods, Sages and Kings*, p. 83.
78. A. Cotterell, op. cit., pp. 18-19; Also see J. Keegan, op. cit., pp. 175-6.
79. Ibid., pp. 1, 67, 69, etc.; J. Keegan, op. cit., p. 176.
80. Ibid., pp. 7-14.
81. Ibid., pp. 235-6.
82. Ibid., pp. 25-9.
83. Ibid., p. 26.
84. Ibid., pp. 32, 179.
85. Ibid., pp. 179, 297.
86. Ibid., p. 55; J. Keegan, op. cit., p. 260.

CHAPTER 2

The Vedic Age

ĀRYANS IN INDIA

It appears that around 1500 BC, a branch of the Āryans migrated to India from Central Asia. They brought in their knowledge of chariots along, manufactured them in large numbers and used them widely in battles against their opponents. The chariots are also belived to have ferried their gods to partake the offerings at sacrificial rituals. In Vedic hymns, Indra is often invoked to come and share the libations, mounting a chariot pulled by trained horses, numbering anywhere between two to seventy.[1]

A branch of these Āryans turned westward from Iran and settled in Western Asia and came to be known as Mitannians. In this connection, Cotterell's view is illuminating:

> It was Indra who, inspired, and actively assisted the Indo-Āryan charioteers on the battlefield. That the chariots they drove were identical to the models used in West Asia points to an invasion of India from that direction: this now seems more than a possibility because of the discovery that the technical terms used in training horses there are very close to Sanskrit, the language of the *Ṛgveda*. . . . It should perhaps be noted, too, how the Mitannians worshipped Indra, Mitra, Varuṇa and the Aśvins, the twin horse-headed gods of ancient India.[2]

But Cotterell's view perhaps needs a little revision. The Indo-Āryans may not have come from western Asia. The Mitannians and Indo-Āryans may have belonged to one and the same family, a branch of which went to settle in Western Asia and the other in India, parting their ways somewhere in Iran. A third branch stayed back in Iran. About the close relationship between Iranians and Vedic Indians, Ragozin has to say the following:

Two such sister nations we have in the Āryan Hindus and Eurasians. It is impossible to do justice to the history and culture of the one without drawing the other into the same field of vision and comparing the two – a process, which necessarily brings out their common origin by presenting identical and similar features, obviously borrowed by neither from the other, but inherited by both from a common ancestry.[3]

This seems to be supported by the archaeological evidence which suggests that proto-Indo-Āryan speakers had become the elite layer of the culture of Bactria and Margiana Archaeological Complex (BMAC) in southern central Asia before spreading to the Indian subcontinent.[4]

Further, the discovery of a cylinder seal depicting a horse-drawn chariot at Tepe Hissar has led Roman Ghirshman to suggest that the proto-Indo-Āryan-speaking elite which ruled the Mitanni Kingdom of northern Syria during 1500-1300 BCE came from north-eastern Iran.[5]

How the early Indo-Āryans fought their battles during the Vedic period is succinctly described by Adolf Kaegi:

When an enemy approaches the Āryan boundaries, earthworks are thrown up, a barricade of timbers created, impassable bulworks of bronze made and sacrifices offered to gods to secure their help. Then, the army advances with loud battle-songs, with the sound of drums and trumpets, with waving banners against the opposing foe. The warrior stands at the left of the chariot, and beside him the charioteer, and the foot soldiers fight in close lines, village beside village, tribe beside tribe. The warrior is protected by brazen coat of mail and helmet; with the bow he hurls against the enemy feathered arrows with poisoned tips of horn or metal and presses on with spear and axe.[6]

The chariot and chariot-rider stand out in this milieu.

VEDIC CHARIOT-MAKERS (*RATHAKĀRAS*)

In the *Ṛgveda* the word *rathakāra*, denoting a professional chariot-maker, is not mentioned. Still, in the Vedic age chariot making had come to be considered an art. It is said that a wright or skilled craftsman made a chariot in the same manner as a singer composed a hymn for the deity.[7] As chariot making was a dignified profession and there was a great demand for the chariots among Vedic Āryans, many craftsmen may have joined the profession.

In the *Ṛgveda* Ṛbhus appear as skilful chariot makers. Their craftsmanship is described in the following hymns:

1. They for the two Nāsatyas wrought a light car moving every way.[8]
2. With cunning (skill), Ṛbhus, rich in treasure, fashioned the two swift tawny steeds who carry Indra.[9]
3. Ye sapient Ones, who made the lightly rolling car out of your mind, by thought, the car that never errs.[10]
4. Come on that chariot which Ṛbhus wrought for you, the chariot, Aśvins that is speedier than thought.[11]

Besides Ṛbhus, some other craftsmen are also mentioned as chariot-makers. A hymn dedicated to Indra says that 'Anus have wrought a chariot for thy courser.'[12] Bhṛgus also appear as chariot-makers in the *Ṛgveda* and they are said to have made a car for Indra.[13] It seems that they subsequently abandoned this profession and joined the priestly class. Vibhvan was another craftsman who made chariot for Varuṇa.[14] Vāja was the expert artificer of gods.[15] In the *Yajurveda*, it is said that the chariot of Indra named Nārāśaṁsa was built by Tvaṣṭṛ.[16] Again there is a reference to a swift chariot made by men who knew their craft.[17] According to Kosambi, 'technology in the *Ṛgveda* amounted mostly to the construction of chariots, tools and weapons of war; it belonged primarily to the God Tvaṣṭṛ and his followers, both seemingly of Indian origin'.[18]

In the *Ṛgveda* artificers like Bhṛgus, Ṛbhus, Vāja and Vibhvan are spoken with respect and many hymns have been dedicated to them. They joined the sacrifice alongside Indra and drank *soma* in his company.[19]

In the *Atharvaveda, rathakāra* appears as a functional caste.[20] In the later Vedic literature, they are categorized differently from the three higher castes.[21] The *Pūrva-Mīmāṁsā Sūtra* says that they belonged to that category of caste called *saudhanavana*, which was neither included in the three upper castes nor in the Śūdra category,[22] but was ranked somewhere in between. This is further confirmed by the *Taittirīya Brāhmaṇa*, when it says that the Brāhmaṇas should consecrate sacred fire in the spring, the Kṣatriyas in summer, the Vaiśyas in autumn and the *rathakāras* in rainy season.[23] In any case, they were entitled to perform *upanayana* (initiation)[24] and commanded respect.[25]

The chariot, if damaged, was repaired as is suggested by a hymn dedicated to Aśvins. 'When he restored to youth and strength Kakṣīvān like a car renewed.'[26] In another context it is said that Aśvins restored Vandanā to health like a car worn out with the length of days.[27] In fact, even two damaged chariots could be made into one and put into use.[28]

THE EARLY CHARIOT

It is widely believed that, the Āryans conquered the subcontinent mounting their chariots and horses, after crossing the high mountains that shield India from the rest of Asia. But this theory was made suspect some fifty years back by S.D. Singh and many others who raised some pertinent questions.

> Did all the Āryans or indeed most of them, the noble and the plebeian, ride across the passes of Khyber on chariots – as they poured into India in their thousands – their wheels with only four or six spokes rumbling along the tortuous paths of the rocky, dumpy and difficult terrain? Was it possible?. . . The primitive chariot held together by belts of ox-hide and cords would fall to bits and pieces on such a rugged highway.[29]

Singh himself answers that,

> The Āryan chariots imported into India must have been dismantled and loaded on horsebck through the Khyber and the difficult mountainous road. Most of our travellers must have ridden their horses or vended their way on foot up and down the hilly tracks.[30]

But was it really necessary to dismantle the chariot and load it on horseback to transport it. It is likely that, this early chariot, made of a lighter wood, might have weighed about 40 kg and could therefore, be transported across the inhospitable terrains physically. The following hymn dedicated to Agni in the *Ṛgveda* seems to support this hypothesis.

Like a swift chariot made by men who knew their art, he with his red limbs lifts himself aloft to heaven (*Ṛ*V, 1.141.8).

Keegan has also tries to find an answer to this:

The rapid dispersion of the chariot ought not to surprise us. . . . Once perfected, the technology of the chariot would have been easy to replicate and even easier to transport and sell; an Egyptian bas-relief of *c*.1170 BC shows a chariot being carried on the shoulders of one man – no feat if it weighed, as a reconstruction did, less than a hundred pounds – and such a highly marketable product would have stimulated production wherever craftsmen with the necessary skills resided.[31]

More recently, Staal has put fourth an explanation and perhaps in a more convincing manner. He states that, 'there was no need for chariots to be carried across mountains. It was the idea of the chariot and precise knowledge how to construct one that were transmitted across the mountains by a handful of experts. That knowledge was enough.'[32] Furthermore, he adds that, 'the tribes who spoke Indo-Āryan imported such chariots into the subcontinent through their oral tradition, that is through their minds.'[33]

Equipped with the knowledge of making chariots, the invaders might have started manufacturing them in different sizes in their new habitat. The reference to a big-sized chariot in the *Ṛgveda* suggests that these were made in smaller sizes as well.[34] The *Vedic Index* sketches the chariot as follows:

Normally there was it seems one pole, on either side of which, the horses were harnessed, a yoke (*yuga*) being laid across their neck; the pole was passed through the hole of the yoke (called *kha* or *tardman*), the yoke and the pole being then tied by the neck (*grīvā*) where the yoke was placed, and also at the shoulder, presumably by traces fastened to a bar of wood at right angles to the pole, or fastened to the end of the pole, if that is to be regarded, as it probably should, as of triangular shape, wide at the foot and coming to a point at the tip. The traces seem to be denoted by *raśmi* and *raśnā*. These words also denote the reins, which were fastened to the bit (perhaps *śiprā*) in the horse's mouth. The driver controlled the horses by reins and urged them on with a whip (*kaśā*). The girth of the horse were called *kakṣya*.[35]

THE DIMENSIONS

The dimensions of the chariot are not detailed in the *Ṛgveda*. But in *Śulbasūtras*, particularly *Āpastamba Śulbasūtra*,[36] *Baudhāyana Śulbasūtra*,[37] *Kātyāyana Śulbasūtra*[38] and *Mānava Śrautasūtra*[39] some details are given and these may hold true in respect of the Vedic chariot as well.[40]

In the *Āpastamba Śulbasūtra*, the dimensions of the chariot are mentioned while stating that the sacrificial altar (*vedī*) for *niruddha paśubandha* sacrifice shall be of the measure of a *ratha*.[41] The back-side of the *vedī* (*paścāttiryak*) should be of the measure of an axle (*rathākṣa*), the eastern side of the measure of a pole (*īśā*) and the front side of the measure of a yoke (*vipatha yuga*).[42] But the actual dimensions of a chariot are mentioned in relation to *caraṇa-ratha* as follows:

i. *īśā* or the pole of a chariot is about 188 *aṅgulas*.[43]
ii. *tiryagakṣa* or the cross axle is about 104 *aṅgulas*.[44]
iii. the *yuga* or the yoke is about 86 *aṅgalas*.[45]

If the dimensions are calculated on the assumption of *Vedic Index* (vol. II, p. 203) that 16 *aṅgulas* (finger breadth) equalled to one foot, the pole length of the chariot would be 11¾ ft, axle length 6½ ft and yoke length 4½ ft. Piggott is, however, inclined to put the width of an *aṅgula* at half an inch for the purpose of calculating the dimentions of a Vedic chariot. As per his calculations the pole length of a Vedic chariot comes to 7¾ ft, cross axle length to 4½ ft and yoke length to 3½ ft. He further argues that the pole length of West Asian chariots ranged between 6 ft and 7½ ft and that the pole of the Vedic chariot may also not have been much bigger than that.[46]

More recently, M. Sparreboom in his study on Vedic chariot has come to a different conclusion as in a recent study on Vedic geometry, Michaels has calculated the value of *aṅgulī* at 1.6 cm. Believing on his calculation Sparreboom says:

> For the chariot this would amount to 300 cm for the pole 166 cm for the axle and 137 cm for the yoke, which would seem perfectly acceptable.

The pole does seem extraordinarily long. But if we consider that the pole may have had a projection in front of the yoke (*praiga*) and possibly extended underneath and even protruded beyond the chariot box at the rear end (as in the Sāñcī chariots) the distance from chariot body to the yoke would be less than 200 cm, which is not too long. A chariot that according to ritual text was drawn by three or four horses is more likely to have been of this large size, rather than of a smaller variety.[47]

In the *Ṛgveda* there are some references to luxury cars (*sukhamaye rathe*). In one hymn, Agni is invoked to ferry the gods in a comfortable chariot.[48] In another hymn Indra is beseeched to join a scarifice mounting a similar chariot.[49] The chariot of Savitar could not have been anything but a luxury car:

> His chariot decked with pearl, of various colours, lofty, with golden pole, the god hath mounted. . . .
> Drawing the gold-yoked car his Bays, white footed have manifested light to all the peoples.[50]

Luxury chariots made of gold or say worked with gold, were thus known.[51] The divine Savitar travelled in his golden chariot, which in addition was decorated with golden ornaments and furnished with golden yoke.[52] The chariots were painted in various colours, such as red (*aruṇa*) and red-brown (*piṣaṅga*).[53] These were further upholstered with some kind of skin to add strength to their body. In the *Atharvaveda*, there is a reference to *vaiyāghra-ratha*, used at the time of consecration ceremony.[54]

It seems that the chariots (light ones) were rested on a platform like a cart called *rathavāhana*, when not in use.[55] A hymn addressed to Indra makes this amply clear:

> Thou hast drunk Soma, Indra, turn thee homeward; thy joy is in thy home, thy gracious consort; where thy tall chariot has a place to rest in and thy strong courser is set free with guerdon.[56]

In the *Yajurveda* also, there is a reference to *rathavāhana* which was called *havi* (*śakaṭa*). It carried not only the chariot but also the armour and weapons of the warrior.[57]

THE COMPONENTS

In the Vedic age, chariot called as *ratha, syandana* and *sthūri* was made of wood is evident from the following hymn:

Lord of the wood, be firm and strong in body: be, bearing us, a brave victorious hero.

Show forth thy strength, compact with straps of leather and let thy rider win all spoils of battle.

Its mighty strength was borrowed from heaven and earth; its conquering force was brought from sovrans of the wood.[58]

The body of the chariot was made of *khayar* (hard wood of *khadira* or *Acacia catechu*) and *śinśapā* (*Dalbergia sisu* also a common tree) wood is indicated by the following hymns dedicated to Indra:

> Enclose thee in the heart of *khayar* timber, in the car wrought of *śinśapā* put firmness.
> Show thyself strong, O Axle, fixed and strengthened:
> throw us not from the car, whereon we travel.[59]

The wheels were, however, made of *kiṁśuka* (*Butea monosperma*) or *śālmali (Bombax ceiba):*

Mount this, all-shaped, gold-hued, with strong wheels, fashioned of *kiṁśuka* and *śālmali*, light rolling.[60]

The *Atharvaveda* says that the chariot made of wood was firmly tied with the ropes made of cow-hide and that these bindings made it as strong as the *Vajra* of Indra.[61] The *Ṛgveda* also refers to a car bound with straps.[62]

The chariot made of wood was treated like a divinity, so much so that it was equated with gods like Indra, Mitra, Varuṇa, etc.:

Thou Bolt of Indra, vanguard of the Maruts, close knit of Varuṇa and child of Mitra. As such accepting gifts which here we offer receive O godlike chariot, these oblations.[63]

The following hymns dedicated to Aśvins also bear this out:

> Aśvins, your swiftly-rolling circumambient car, which he who worships must invoke at eve and dawn (*ṚV*, 10.39.1).
> That circumambient car, worthy of sacrifice, we call with our pure hymns at earliest flush of dawn (*ṚV*, 10.41.1).

Explaining *rathavāhanam* occurring in a hymn, Griffith says that the word seems in this place to mean also the oblation offered by the warrior to the ideal war chariot personified or to a tutelary diety of chariots.[64]

In different parts of the world different kinds of wood were used for making chariots.

> Analysis of surviving Egyptian chariots show that the body, yoke, wheel hub and felloe were cut from elm, which would have been imported from Ugarit. The draught pole was made of willow, and the wheel spokes of plum. Neither of these trees grew in Egypt, while the nearest source of the birch bark used for coverings was Asia Minor.[65]

Indeed these woods were imported by the Egyptian Pharaohs to make chariots.

THE HORSE (*AŚVA*)

According to Frits Staal, there is no doubt that horses were introduced in India from elsewhere as no evidence of their existence is met in the remains of Indus Valley Civilization. Their earliest traces in South Asia come from Pirak, south-east of the Bolan Pass and which dates from the seventeenth century BC.[66] Elaborating further he says that horses have always been rare in India. They do not thrive on Indian soil and the climate does not smile on them. But they are rampant in the Vedas. In the *Ṛgveda*, there are 792 occurrences of the word *aśva* in various forms.[67]

Generally, two horses were yoked to the chariot and were bound to it by the neck, by the flanks and by the mouth.[68]

i. And thy two highly lauded bays, strong stallions, draw thy car.[69]
ii. Let the two long-manned bay steeds bring Indra to drink the *soma* juice.[70]
iii. Thy two bays with peacock tails convey thee hither, steeds with their white backs.[71]
iv. On both sides of the car they yoke the two bay coursers dear to him.[72]
v. Harness thy pair of strong bay steeds.[73]
vi. Two blustering steeds of the wind god thunderer, that speed along the shining paths.[74]

In the *Ṛgveda*, there are also references to more than two horses being yoked to a chariot. The chariot of Agni was pulled by four horses.[75] Seven horses were yoked to the chariot of Sūrya.[76] There is also a reference to forty bay horses yoked to ten chariots, leading a long procession.[77]

In some cases the number of horses yoked to a chariot has been highly exaggerated. In a hymn, Vāyu is invoked to join a feast along with Indra, mounting a chariot pulled by 99, 100 or 1,000 steeds.[78] In another hymn, Indra is invited to come by a chariot pulled by 1,000 horses.[79] In yet another hymn, Indra is invoked to share the libation mounting a chariot pulled by 4, 6, 8, 10, 20, 30, 40, 50, 80, 90, or 100 horses.[80] The exaggerated number of horses perhaps symbolized the power of the deity, they were associated with. Occasionally, the chariot was pulled by one horse only is suggested by the following hymn. 'Between both poles the car-horse goes pressed closely, as in his dwelling moves the doubly wedded.'[81] There is also a reference to three bays (chestnut coloured horses) harnessed to a chariot.[82]

Besides the horses, the mares, stallions, mules and asses were also yoked to the chariots. Viśvedevas yoked red mares to their car,[83] and so did the Maruts.[84] The chariot of Indra was pulled by strong stallions.[85] Nāsatyas mounted a car drawn by stallion asses.[86] In *Chāndogya Upaniṣad*, there is a reference to mule drawn chariot.[87] In the *Vādhūla Śrautasūtra*, it said that four mule-driven chariots should be sent to sixteen *janapadas* of rival kings to subdue them.[88] Interestingly, at one place the chariot of Maruts is stated to have been pulled by deers.[89] There is also a reference to a chariot being pulled by bull and oxen.[90]

Horses of different colours pulled the chariots of different gods. The mares yoked to the chariot of Viśvedevas were of red colour.[91] Bay steeds were harnessed to the chariot of Sūrya.[92] Two red steeds and two ruddy steeds pulled the chariot of Agni.[93] Uṣā dawned on a chariot drawn by purple horses.[94] Indra has been called lord of tawny steeds.[95] The chariot of Bṛhaspati was pulled by red horses.[96] Two horses, one red and the other purple, were yoked to the chariot of Vāyu.[97] The colours of the horses perhaps carried some symbolic meaning. It seems that horses were trained before being yoked to the chariot.[98]

ACCOUTREMENT OF HORSE

The accoutrement associated with the chariot-horse included yoke, whip, reins, straps and harness. A hymn devoted to Indra reads:

> The rich new car hath been equipped at morning: four yokes it hath, three whips, seven reins to guide it.[99]

In a hymn devoted to Aśvins, the whip is described as 'droping with honey'.[100] This is suggestive of a soft whip, which did not strike the animal hard. Another hymn addressed to whip runs as follows:

> He lays his blows upon their backs, he deals his blows upon their thighs. Thou, Whip, who urgest horses, drive sagacious horses in the fray.[101]

The rein was intended to direct the horse.[102] It seems that, due care was taken in making the rein as a hymn dedicated to Maruts says that, 'your reins be fashioned well'.[103] In the *Yajurveda*, it is said that the bridle added to the decor of the horse and the chariot.[104]

The harness of a Vedic draft horse has been discussed in some detail by R.L. Mitra:

> In olden times (it) included a body-roller, a collar, and a bridle. The body roller was apparently plain and tied where in our times the sursingle (strap for holding a saddle) is buckled; but without any padding or cloth underneath; differing in this respect from the Assyrian harness, which always included a rich saddle cloth. It was intended to prevent the traces (strap attached to horses collar) from hanging low when the horse was checked or backed.[105]
>
> The collar was light below, but heavy at top. . . . In fact, the idea of a collar was derived from a bullocks hump, and the contrivance was designed with a view to give a false hump to the horse and the traces were so adjusted is to throw the weight much higher up than the point, which bears the greatest strain under an ordinary collar.[106]

THE WHEEL IN VEDIC LITERATURE

Wheel made of wood was another vital component of the chariot:

> Mount this, all-shaped, gold hued, with strong wheels, fashioned of *kiṁśuka* and *śālmali*, light rolling, bound for the world of life immortal. Sūryā: Make for thy lord a happy bridal journey.[107]

Generally, the chariot was a two-wheeled vehicle. In a Vedic hymn, Aśvins are implored to 'come like two chariot wheels at dawn'.[108] Another hymn addressed to Maruts says that, 'your axle turns together both the chariot wheels'.[109] There are also references to three-wheeled chariots, particularly in relation to Aśvins.[110] Viśvedevas travelled in a seven-wheeled chariot[111] and so did the Soma and Pūṣan.[112] The vehicle of the sun was, however, a one-wheeled chariot.[113]

In the *Ṛgveda* many components of the wheel are mentioned. These include spokes (*ara*), felly (*nemi*), rim (*pavi*), nave (*nābhya*), linch-pin (*aṇi*), hollow aperture (*kha*) and so on. Spokes (*ara*) were a vital constituent of the wheel and it was only after their invention that the chariot became an effective war machine. A hymn dedicated to Maruts is significant in this regard:

> Pay honour to these Maruts and sing praise to them, for of the wheel-spokes of the car of these loud roarers, none is lost: this is their power, this moves them to give mighty gifts.[114]

Two more hymns denote the importance of spokes:

> For as the felly holds the spokes, thou (Agni) with thy might pervading hast been born encompassing them round.[115]
> (Maruts) like spokes of car-wheels in one nave united, ever victorious like heavely Heroes.[116]

The number of spokes, in a wheel is not specified in the *Ṛgveda*. Wilson, believes that the number of spokes in each wheel might have been five.[117] Perhaps, with the passage of time more spokes were added as there is a reference to a wheel with twelve spokes.[118] According to Singh, Vedic chariot might have had four to eight spokes and that more might have been added later.[119]

Felly (*pradhi)*, which firmly held the spokes, was another vital part of the wheel.[120] In a hymn dedicated to Maruts, it is said that the fellies of their chariot-wheel were provided with knives.[121] Griffith explains that 'their war chariots have sharp scythe-like blades attached to their wheels, or sharp edges on their fellies'.[122] This projection of scythe blades from the chariot wheels might have rendered an attack from the flanks hazardous to the attacker.

In fifth century BC Ajātaśatru, the ruler of Magadha, had invented a mechanical device called *ratha-mausala* to provide protection to his chariots[123] or to damage enemy chariots. It seems that chariot-fellies left a mark on the ground they treaded. In a reference to Maruts it is said that, 'Earth they have smitten with the chariot's felly'.[124] No doubt the fellies were made strong.[125]

Nave (*nābhya*) was the hub or the central part of a wheel through which the axle passed. The wheel was secured with a rim (*nemi* or *pavi*) to provide it strength and smoothness. Linchpin (*āni*) and a hollow aperture (*kha*) were also part of the wheel.[126]

Box (*kośa*) was that part of the chariot which provided space both to the warrior and the charioteer to sit or stand safely. It was perhaps made of wickerwork as in the Aegean and Celtic chariots, or of leather on a light wooden framework as in Egypt.[127] It was fastened to the wooden axle (*akṣa*) with straps of cowhide.[128]

Axle (*akṣa*) was another vital part of the chariot, which carried the box of the car on top and the wheels on either side. It was made of *araḍu* wood (*Calosanthes indica*)[129] of golden hue.[130] A hymn prays 'show thyself strong, O Axle, fixed and strengthened: throw us not from the car whereon we travel'.[131] The wheels might have been secured to the axle with some kind of nails.

Garta or *vandhura* was the seat of the chariot. Generally, it provided space for two persons, viz. the warrior and the charioteer.[132] However, the chariot of Aśvins provided sitting space for three persons,[133] viz., the charioteer and the two Aśvin brothers. There is also a reference to eight-seater chariot, which was perhaps a transport vehicle.[134] In the *Atharvaveda*, the warrior has been called *savyaṣṭhā*[135] because he occupied a seat on the left side of the chariot.

Car-pole (*īṣā*) was a vital component which bound the body of the chariot to the yoke (*yuga*). It was passed through the hole of the yoke and then tied with cords. A hymn in the *Ṛgveda* says:

> Bind to the pole with cords of holy order the long-manned ruddy steeds, who sprinkle fatness.[136]

The cords were, indeed, used so that the pole did not slip.[137]

Yoke (*yuga*) was that component of the chariot to which the horses were tied. There are many references to horses being yoked to the chariot in the *Ṛgveda.*

i. Whose (Indra) pair of towney horses yoked in battles-foemen challenge not.[138]
ii. On both sides of the car they yoke the two bay coursers dear to him (Indra).[139]
iii. Now, Indra, yoke thy two bay steeds.[140]
iv. When Indra thou dost yoke thy steeds there is no better charioteer.[141]

THE CHARIOTEER (*SĀRATHI*)[142]

It seems that the office of the charioteer did not exist in the early Vedic age. The warrior (*rathin*) himself drove the chariot to the battlefield, brought it to a halt and then joined the battle. *Raśmi* and *raśanā* denoted the 'reins' (perhaps *śiprā*), which was fastened to the bit in the horse's mouth. The charioteer controlled the horses by reins and urged them on with a whip (*kaśā*).[143] When at a distance, the charioteer might have shot the arrows and hurled the javelins or spears targeting the enemy. But when in the thick of the battle, he might have fought with a sword, probably made of copper, as iron came into use later. That the warrior and charioteer was one and the same person is evident from the following hymns:

> With holy prayer, I yoke thy long-maned pair of bays: come hitherward; thou holdest them in both thy hands.[144]
> When Indra thou dost yoke thy steeds, there is no better charioteer.[145]
> Ascend this chariot, thou whose hand yields thunder, and draw the rein, lord of noble horses.[146]
> Best of all charioteer is he.[147]
> Indra sitting in his chariot held the bridle of his trained horses.[148]
> Standing on his chariot, the excellent charioteer leads the horses wherever he wishes. Praise the power of reins: the ropes follow his mind.[149]

However, with the passage of time, Vedic Āryans realized that the task of managing the horse while fighting the battle compromised the efficiency of the warrior. The office of the charioteer was, therefore, created to manoeuvre the chariot in the battlefield, so that the warrior was left free to conduct the battle. Many hymns devoted to Indra in the *Ṛgveda* prove this beyond any doubt:

i. Those charioteers of thine, best skilled to draw the rein.[150]
ii. Indra made room for his car driver Kutsa, who sate beside him.[151]
iii. Drivers stand on his golden chariot firmly stationed.[152]
iv. Whom two invoke on one chariot mounted, each for himself.[153]
v. Upstanding in the car the skilful charioteer guides his strong horses.[154]
vi. We hold on to Pūṣan and Indra as the charioteer holds the bridle.[155]

In the *Ṛgveda*, Pūṣan has been described as a great charioteer, who accompanied Indra in his exploits:

> And he is best of charioteers. Indra, the hero's Lord, allied
> With him as Friend, destroys the foes.[156]
> We pray for wealth to thee (Pūṣan) most skilled of charioteers, with braided hair,
> Lord of great riches, and our Friend.[157]
> As one who drives a car draws in his reins, may he draw Pūṣan near,
> And Indra, for our great success.[158]

It is notable that in some hymns, Kutsa also figures as a charioteer of Indra:

- Once to the driver of his chariot, Kutsa he gave up greedy Suṣṇa.[159]
- Let the steeds bring you both, Indra and Kutsa, borne on the chariot within hearing distance.[160]
- Indra made room for his car driver Kutsa who sate beside him, when he gained the sunlight.[161]

In the *Atharvaveda*, Candramā also figures as a *sārathi* of Indra.[162] The institution of charioteer had thus, come into existence in the Vedic age, albeit a little late. The commander occasionally engaged more than one charioteer to manage the horses, if the number of horses yoked to the chariot exceeded two. Pūṣan was, indeed, the greatest charioteer of the age. In view of this abounding evidence on charioteer in the *Ṛgveda*, it is difficult to agree with Dikshitar that the warrior in his chariot was his own driver until the function of the two were separated about the time of the *Aitareya Brāhmaṇa*.[163]

However, occasionally the warrior himself served as charioteer. In the *Yajurveda*, Indra is described as yoking his horses and holding the reins.[164]

The first and foremost task of a charioteer was to manage the horses. How he did it is best described in the following hymns:

> He lays his blows upon their backs, he deals his blows upon their thighs. Thou whip (*aśvajāni*), who urgest horses drive sagacious horses in the fray.[165]
>
> A good charioteer sitting in the chariot directs the horses wherever he desires to.[166]
>
> The warrior hits the horses on the hips and the body to lead them onward.[167]
>
> Upstanding in the car the skilful charioteer guides his strong Horses on whithersoe'er he will.
>
> See and admire the strength of those controlling reins which from behind declare the will of him who drives.[168]

Subsequently, a goad (*pratoḍa*) also came into vogue to discipline the horses.[169]

It may be mentioned here that charioteers enjoyed a good status in the society during the Vedic times. In the *Ṛgveda* chariot artificers like Ṛbhus, Vāja and Vibhavan could even drink *soma* in the company of Indra.[170] In some Brāhmaṇas, the *sūta* (charioteer) and *takṣa-rathakārau* (carpenter and chariot maker) are included in the category of *ratnins* of the state.[171]

THE CHARIOT-WARRIOR (*RATHIN*)

The warrior (*rathin*) was key to the chariot ensemble. His dress, accoutrements and the weapons are occasionally mentioned in the *Ṛgveda*:

> The warrior's look is like a thunderous rain-cloud's, when, armed with mail he seeks the lap of battle.
> Be thou victorious with unwounded body: so let the thickness of thy mail protect thee.[172]
> Thy vital parts I cover with thine armour: with immortality King Soma clothe thee.[173]
> It (brace) compasses the arm with serpent windings, tending away the friction of the bowstring.[174]
> Visors wrought of gold are laid upon your head.[175]

Similarly, adornments and equipment of a chariot warrior can also be gleaned from the following hymns dedicated to Maruts:

> Who shine self-luminous with ornaments and swords, with breastplates, armlets and the wreaths.[176]
> Lances are on your shoulders, anklets on your feet, gold chains are on your breasts, gems, Maruts on your car.
> Lightnings aglow with flame are flashing in your hands.[177]
> Armed with your daggers, full of wisdom, armed with spears, armed with your quivers, armed with arrows, with good bows.
> Good horses and good cars have ye, O Pṛṣni's son: ye Maruts, with good weapons go to victory.[178]

Bow was the principal weapon of a chariot warrior. 'With bow let us win kine, with bow the battle' says a hymn.[179] Another hymn describes the arrow 'Her tooth a deer, dressed in an eagle's feathers, bound with a cowhide, launched forth, she flieth onward'.[180] Griffith explains that the point of the arrow was made of a piece of deer's horn attached to the shaft with leather strings and the butt of the arrow was feathered.[181] The arrow-head made of iron was smeared with venom.[182] The warrior carried a quiver on the back. 'Slung on the back, pouring his brood, the quiver vanquishes all opposing bands and armies'.[183] The whip was used to manage the horses in the battlefield.[184]

In a chariot, the *rathin* took his position on the left side.[185] The *sārathī*, indeed took his position on his right. The space provided in the chariot for the warrior to sit was called *garta* or *vandhura.*[186] In this connections, the following observation made by Piggott is significant.

The chariot held two people, the warrior and his driver. The warrior was on the left, and seems to have been provided with a seat, which he could use at least when he was not actively engaged in warfare. . . . The charioteer had no seat provideed. One of his title, *sthatr*, he who stands, emphasizes the distinction between him and the bowman.[187]

Perhaps, a standing charioteer could have better control over the horses and could adjust better to the twists and turns of battle unfolding before his eyes.

It follows that a Vedic chariot-warrior put on a mail to protect his body, a visor to protect his face and head, and a brace to protect his hand. He adorned his person with ornaments like armlets, wreaths, anklets, gold chains, gems, etc. The weapons carried by him included bow and arrow, dagger, sword and lance. A quiver was hung on the back.

It seems that the war-chariots in the Vedic age were accompanied by guards (*pālayitāraḥ*), apparently those who attended the chariot of the chief. Wilson, following Sāyaṇa translates the relevant hymn as follows:

The guards (of the chariot), revelling in the savoury (spoil), distributors of food, protectors in calamity, armed with spears, resolute, beautifully arranged, strong in arrows, invincible, of heroic volour, robust and conquerors of numerous hosts.[188]

It is likely that all chariots engaged in war were protected by guards. In ancient China more than seventy guards accompanied a chariot. In the *Mahāhbārata* also the chariot of the commander was invariably protected by important warriors on all sides.

CHARIOT IN BATTLE

The chariot dominated the Indian battlefield during the Vedic age and prayers were offered to Indra to grant victories to chariot warriors.[189] The following hymns in the *Ṛgveda* amply substantiate this:

i. And fain to come forth first amid these armies, this way or that with rows of cars he (*dadhikrās*) rushes.[190]

ii. What time the warrior Indra goes to battle, borne by noble steeds.
Best of all charioteers is he.[191]
iii. Exceedingly strong in war he (Indra) stays the chariot wheel.[192]
iv. With all thy chariot's force assail the man who shows himself thy foe.[193]
v. Indra, do thou protect our car that mingles foremost in the fights.[194]
vi. With swiftly-racing chariot may we conquer, and rich and blessed be our gains in battle.[195]
vii. The lofty battle-car thou broughtest forward.[196]
viii. As such true hero, for great joy of battle mount thy terrific car, O brave and manly.[197]
ix. Indra ascend thy car to smite down Vṛtra.[198]
x. Good horses and good cars have ye, O Pṛṣṇi's sons: ye Maruts, with good weapons go to victory.[199]
xii. King, on thy chariot-sieve thou goest up to war, and with a thousand weapons winnest lofty fame.[200]
xiii. O, Bṛhaspati, you be the destroyer of *rākṣasas*. Moving fast in thy chariot chastizing and destroying the enemy armies conquering the evil doers in battles, you be the protector of our chariot.[201]

There are many more references to chariot warriors joining the battle. But it is notable that the adversaries of Indra are nowhere reported as fighting mounted on chariots in the *Ṛgveda*. It implies that the *asuras* and *rākṣasas*, the natives who fought against the Āryan invaders, were not familiar with chariots. The adversaries might have followed different forms of battle tactics.

In the *Atharvaveda*, which was a later composition, there is a reference to an encounter between two armies of chariot warriors. A prayer seeks a boon that the enemy chariot warriors be killed in battle and their bodies be devoured by vultures.[202] This appears to be a reference to a duel between two Vedic Āryan tribes. But if the reference is to a battle between the Āryans and a native tribe it is likely that by now the natives had also learnt the use of chariot in battle.

The Vedic war chariots raced at a very fast pace. In a hymn the Aśvins are implored to come on their chariot, which moved faster than thought.[203] In another context the speed of a chariot has been equated with an arrow flight.[204]

But the tracks could not always be evenly laid for the chariots to race on. The charioteers at times found it difficult to negotiate them:

> When Indra in the mighty fray thou urgest chargers to their speed. . . .
> Speeding like rivers, rushing down a steep descend, responsive to the urging call . . . held in by reins in both the drivers hands.[205]
> As drivers of the cars avoid ill road, let sorrows pass us by.[206]

But the Vedas are silent on the battle formations of chariot corps. The basic unit might have been a chariot mounted by a *rathin* and driven by a *sārathi.* Each of these chariots might have been attached to a bigger formation.

The earliest reference to a chariot formation is met in the *Vādhūla Śrautasūtra.* For the protection of Aśvamedha horse four chariots teams of different composition were formed as follows:

1. 300 royal princes in bronze mail armed for battle accompanied by armed charioteers driving chariots with overshields and yoked with four horses.
2. 300 non-royal warriors armed for battle accompanied by charioteers unarmed for battle, driving chariots covered with overshilds and yoked with three horses.
3. 300 headmen not armed for battle, driving chariots not covered with overshields and yoked with two horses.
4. 300 charioteers and villagers, driving chariots yoked with one horse on uneven road.

The four chariot groups were organized in four batches of 75 persons each and deployed on four sides of the horse.[207]

In contempary Egypt, ten chariots formed a troop under the first charioteer (*kedjen-tepy*). Five such troops formed a squadron under a standard bearer of the chariot warrior. Bellicose titles were given to these units which suggest that they were basic tactical and administrative units. Several squadrons could be combined to from a *pedjet* (host) led by a *hery-pedget* (host-chief).[208] It is likely that the Vedic Āryans also organized their chariot corps somewhat like the Egyptians.

Interestingly, a chariot represented in a painting dated in the reign of Thutmose III of Egypt (1495 BC) is stated to be a present from a people called *rat-n-no* or *rathinas* (charioteers). These *rathinas* have been identified by Henry Torrans with the Vedic Āryans of Punjab.[209] But there is a fundamental difference between the said chariot and the chariot as described in the *Ṛgveda.* The former had only one pole with the yoke at the end whereas the latter had three shafts (*triveṇu*) to each car and, therefore, these are described as traiangular. Thus the said chariot could not have been a present from the Vedic Āryans of Punjab.[210]

CHARIOT-RACE

Besides the use in warfare, chariots were also used in racing, which was a favourite sport. Racing was indulged in for amusement and to earn prizes. The race took place along a course to a mark round which the chariots turned and came back again – the *aikavartana* of the Mitannian guide to the turf.[211]

Among the Boghazkoi Keui documents of fourteenth century BC there is a fragmentary handbook on chariot racing, written by a Mitannian named Kikkuli. Interestingly, among the technical terms used for so many 'turnings' round the course he employs forms which are in close proximity to Sanskrit – *aikarvartana, teravartana, panzavartana, shattavartana* – for one, three, five or seven laps of the race (Sanskrit *vartanam,* a turning).[212]

According to A.C. Das,

> Ṛgvedic Āryans were exceedingly fond of horse-racing and chariot-racing, not only for the sake of the fun and the excitement they afforded, but also for exercising the horses, the riders and chariot drivers, and keeping them always fit and efficient. Every village must have possessed its own race-course where the horses were regularly exercised, and where races were run on special festive occasions, and prizes given away to the winners.[213]

Adolf Kaegi says that chariot race was 'the peaceful preparation for the decisive struggle on the battlefield'.[214]

A hymn in the *Ṛgveda* proves it beyond any doubt that chariot races were organized and that charioteers prayed to Indra and other gods for victory:

O Indra, help our chariot on, yea, Thunderer, though it lag behind:

Give this my car the foremost place
Go there! Why sittest thou at ease?
Make thou my chariot to be first:
And bring the fame of victory near.
Assist our car that seeks the prize. What can be easier for thee?
So make thou us victorious.[215]

In another hymn, Indra is again implored to assist in securing the victory in race:

> With hymn and coursers we will gain, Indra, through thee, both steeds and spoil.
> Most glorious, and the proffered prize.
> Thou Indra, Lover of the Song, when men must stir to help, hast been
> Great in the contest for the prize.
> Slayer of foes, whatever aid of thine imparts the swiftest course,
> With that impel our car to speed.
> As skilfullest of those who drive the chariot, with our art and aim,
> O conqueror, win the proffered prize.[216]

The following hymns also refers to the chariot race:

> As for a chariot-race, the skilful Speaker; Chief, Sage, Inventor, hath, with song, been started.
> The sisters ten upon the fleecy summit drive on the car-horse to the resting places.[217]
> Thou conquerest thus, with might when car meets car, and when the prize is staked.[218]

As the main objective of a participant in a chariot race was to win the prize the gods were repeatedly invoked:

- Such as thou (Indra) wast of old, and art now to be called on when the prize Lies ready, listen to our call (*Ṛv*, 6.45.11).
- (Pūṣan) clear paths that we may win prize (*Ṛv*, 6.53.4).
- Thou (Soma Pavamāna) conquerest thus with might when car meets car, and when the prize is staked. (*Ṛv*, 9.53.2).

Commenting on the chariot race R.S. Sharma says that 'of such tests as imply the election of the chief at some early stage, the chariot race occupies an important place. It forms a part of the Vājapeya sacrifice. . . . In the race the royal sacrificer competes with sixteen others. The total number of the chariots being seventeen (*Śatapatha Br.* 5.1.5.6-10). The chariots start with the beating of drums and raising of war cries. A *rājanya* shoots an arrow for fixing the goal of the race.[219]

Sharma adds that,

the chariot race, the product of a developed social stage seems to have been a later version of an older test for proving the competitors superiority in valour and physical power. It was intended to detect the military qualities of the candidate for the post of the king or the chief of the tribe. The victory cries made in connection with this rite make it clear that it was the reflection of the actual practice in military campaigns.[220]

He has rightly espoused the probability that

the ceremony of chariot race was inherited by Vedic Āryans from the original stock of the Indo-Europeans, who used this method in the selection of their chief. In India, also the race was undoubtedly used for this purpose for at the very beginning of the Vājapeya sacrifice, it is stated that the kingdom belongs to him who wins the race.[221]

During the Vedic age, chariot completely dominated the Indian battlefield. The Āryans, who introduced this war machine in India owed their victories to chariot and chariot alone. The natives opposed them with an army of foot soldiers as elephant and horse had not then evolved as effective instruments of war. Naturally, they were easily defeated. It is likely that elephants and horses were also introduced in the Indian battlefield by the Āryans after they had settled in India.

VEDIC CHARIOT BATTLES

There are many references to chariot battles in the *Ṛgveda*. To illustrate the point some passages from the book are quoted below:

> Thou has unclosed the prisons of the waters; thou hast in the mountain seized the treasure rich in gifts.
> When thou hadst slain with might the dragon Vṛtra, thou, Indra, didst raise the sun in heaven for all to see.
> With wondrous might thou blewest enchanter fiends away, with powers celestial those who called on thee in jest.
> Thou hero hearted, hast broken down Pipru's forts, and helped Ṛjiṣvan when the Dasyus were struck dead.
> Thou savedst Kutsa when Śuṣṇa was smitten down; to Atithigva gavest Sambara for a prey.
> E'en mighty Arbuda thou troddest under foot: thou from the old wast born to strike the Dasyus dead.[222]

In the said hymns, Indra is described as the conqueror of many demons like, Śambara Vṛtra, Śuṣṇa, Pipru, Kutsa, Arbuda, etc. He assisted Ṛjiṣvan in his battle against the Dasyus (demons) and killed many of them. He also killed Vṛtra, released waters and made the sun shine in the sky.

> These our libations strength-inspiring, Soma draughts, gladdened thee in the fight with Vṛtra, Hero Lord,
> What time thou slewest for the singer with trimmed grass ten thousand Vṛtras, thou resistless in thy might.
> Thou goest on from fight to fight intrepidly, destroying castle after castle here with strength.
> Thou Indra with thy friend who makes the foe bow down, slewests from far away the guileful Namuci.
> Thou hast struck down in death Karañja, Parṇaya, in Atithigva's very glorious going forth.
> Unyielding, when Ṛjiṣvan compassed them with siege, thou hast destroyed hundred forts of Vaṅgṛda.
> With all out-stripping chariot-wheel, O Indra, thou far-famed has overthrown the twice ten Kings of men.
> With sixty thousand nine-and-ninety followers who came in arms to fight with friendless Suśravas.[223]

The hymns state that the Asuras, Vṛtra, Namuci, Karañja Parṇaya and Vaṅgṛda were vanquished by Indra mounting his chariot. In the process, he also destroyed a large number of enemy forts by way of assisting Ṛjiṣvan. He also vanquished a 60,099 strong army, which had taken field against Suśravas. Many more hymns dedicated to Indra refer to his military exploits.

> For Kutsa, with thy thousand, thou at day-break didst hurl down greedy Śuṣṇa, foe of harvest.
> Quickly with Kutsa's friend destroy the Dasyus, and roll the chariot wheel of Sūrya near us.
> Thou to the son of Vidathin, Ṛjiṣvan, gavest up mighty Mṛgaya and Pipru. Thou smotest down the swarthy fifty thousand, and rentest forts as age consumes a garment.[224]

Indra with his army of one thousand vanquished Śuṣṇa, a demon, to protect the sage Kutsa. He got hold of mighty demons Mṛgaya and Pipru and gave them over to Ṛjiṣvan. Further, Indra destroyed fifty thousand hostiles and brought down their forts to dust.

In the following hymns, dedicated to Viṣṇu, a fierce battle is referred to:

> O Heroes, ye have conquered in your battles even the bull-jawed Dāsas wiles and magic.
> Ye have, destroyed, thou, Indra, and thou, Viṣṇu, Śambara's nine-and-ninety fenced castles.
> Ye twain smote down a hundred times a thousand resistless heroes of the royal Varcin.[225]

Viṣṇu and Indra in a joint battle-effort conquered the Dāsas. They also destroyed 99 castles of the hostiles and killed one lakh Varcin's soldiers in the encounter.

It appears that Āryan immigrants enjoyed a distinct advantage over the natives in the battlefield by virtue of their chariot arm. The natives fighting on foot could not match their might indeed. The assumption is substantiated by a hymn in the *Atharvaveda* which says that a chariot warrior easily conquers a foot-soldier.[226] In another hymn a prayer is made to Indra, Agni and Kāma to mount the chariots and humiliate the enemy.[227]

BATTLE OF PARUṢṆI

Of all the battles described in the *Ṛgveda*, the battle of Paruṣṇi, also called the battle of ten kings, appears to be the most significant. The details of the battle are, however, spread over the whole book and need collation. However, hymn 18 in the book seven provides some useful details.[228]

The battle was fought between King Sudās, the grandson of Deodāsa and son of Pijavana of the Tṛtsu tribe, and a confederacy of ten kings led by Bharatas. Tṛtsus lived along the banks of Paruṣṇī (Rāvī), while Bharatas inhabited the upper reaches of rivers Sarasvatī, Dṛṣadvatī and Āpayā.

It seems that extensive conquests of Sudās had brought many tribes under his rule. This might have fanned jealousy and discontent. Consequently, Bharatas, Anus, Druhyus, Yadus, Turvaśas, Purus, Śimyus, Ajas, Śigrus and the Yakṣus joined hands to defeat Sudās. Bheda, another tribal chief also joined in. It is said that Ṛṣi Viśvāmitra, who had been expelled from the court of Sudās by Ṛṣi Vasiṣṭha and had now joined the Bharatas was instrumental in the creation of this confederacy against Sudās. It seems that Sudās also roped in some northern Āryan tribes on his side, perhaps under the advice of his preceptor Vasiṣṭha.

Sudās arrayed his army on the northern side of River Paruṣṇī. The confederates on the other hand negotiated rivers, Śutudrī and Vipāsa (Beas) in their upper courses to encamp on the southern bank of Paruṣṇi, with loaded wagons and chariots. (*Ṛv* 3.33.9). When the opposing armies had arrayed on either side of the river the confederates planned a stratagem to gain an easy victory. They decided to effect a breach on the embankment of river Paruṣṇī to flood the Tṛtsu territory. But Sudās foiled the attempt with his bold initiative and killed the person entrusted with the task.

Sudās, on the other hand, adopted an outflanking strategy to defeat the confederates. He picked up a strong contingent of his army, crossed the River Paruṣṇī higher up, unseen by the confederate army, perhaps during the night, and launched a fierce attack from behind. Trapped between the river and the attacking army of Sudās, the confederate army suffered heavy casualties. Twenty-one leaders were slain and many drowned. Those who managed to swin across to the northern bank were killed or captured by the encamped army of Sudās.

After earning this major victory against the confederates Sudās went about invading the territories of collaborators. An allusion to these encounters is made in the following hymns dedicated to Indra (*Ṛv* 7.18):

The king who scattered one-and-twenty people of both Vaikarṇa tribes through lust for glory.
Thou thunder-armed o'erwhelmedst in the waters famed ancient Kavaṣa and then the Druhyu.
Indra at once with conquering might demolished all their strong places and their seven castles.
The goods of Anu's son he gave to Tṛtsu.
The Anavas and Druhyus, seeking booty, have slept, the sixty hundred, year, six thousand and six-and-sixty heroes (6666).
These Tṛtsus under Indra's careful guidance came speeding like loosed waters rushing downwards.
The foemen, measuring exceeding closely, abandoned to Sudās all their provisions.
To thee have all thine enemies submitted: e'en the fierce Bheda hast thou made thy subject.
There he stripped Bheda bare of all his treasures.
The Ajas and the Śigrus and the Yakṣus brought in to him as tribute heads of horses.

Thus, Sudās assisted by Indra carried a successful campaign against his adversaries. Some of them died fighting while some others choose to submit. In these battles, many people were killed, many forts and cities demolished, many territories conquered and many treasures grabbed by victorious Sudās.

According to David Frawley,

most central to the historical interpretation of the *Ṛgveda* is the story of Sudās and the legendary battle of ten kings (*dasarājña*). . . . The great war of his time was an event in early ancient Indian history probably as significent as the Mahābhārata war was in later ancient times. It similarly may have involved most of India.[229]

A MEANS OF TRANSPORT

In the Vedic times, chariot was widely used as a means of transport. The following hymns in the *Ṛgveda* amply prove this:

Harness thy pair of strong bay steeds, long maned, whose bodies fill the girths.

And, Indra, Soma Drinker, come to listen to our songs of praise.[230]
Let the swift steeds who carry thee, though-yoked and dropping holy oil.
Bring the gods to the Soma draught.[231]
Harness the Red Mares to thy car, the Bays, O God, the flaming ones:
With those bring hitherward the Gods.[232]
Vāyu let fleet-foot coursers bring thee speedily to this our feast, to drink first of the juice we pour.[233]
Come on your chariot, ye who travel widely, come to this sacrifice of ours, Nāsatyas.[234]
Let thy Bay steeds bring thee, the strong, hither to drink the Soma draught . . . hither let the Bay coursers bring Indra upon his easiest car.[235]

In this hymn, the term *sukhatame rathe* need not perhaps be translated as easiest car. If might have denoted a comfortable or luxury car, which was used by Indra on festive occasions. In the *Ṛgveda* the Aśvins are also described as 'the Gods borne in a noble car'.[236]

There are also references to chariots made of gold or decorated with gold in the *Ṛgveda.* The following hymn dedicated to Aśvins illustrates this.

Ascend your car with golden seat, O Aśvins,
And with reins of gold,
That reaches even to the sky.
Golden is its supporting shaft, the
Axle also is of gold,
And both the wheels are made of gold.[237]

The chariot of Savitṛ is similarly described as made of gold:

Borne in his golden chariot he cometh. . . .
His chariot decked with pearl, of
Various colours, lofty, with golden pole.
Drawing the golden yoked car his Bays
White footed.[238]

The car of Indra was also made of gold:

A thousand and hundred steeds are harnessed to thy golden car.[239]
Yoked to thy chariot wrought of gold.[240]
He stays his golden car, yoked with Bay horses.[241]

DIVINE ATTRIBUTES

By virtue of its importance in the life of the Āryans, over the times, the chariot came to assume an aura of divinity. Many hymns in the *Ṛgveda* substantiate the assumption:

Lord of the wood, be firm and strong in body: be, bearing us, a brave victorious hero. Show forth thy strength, compact with straps of leather, and let thy rider win all spoils of battle.[242]
Its mighty strength was borrowed from the heaven and earth: its conquering force was brought from sovrans of the wood.
Honour with holy gifts the Car like Indra's bolt, the Car bound round with straps, the vigour of the floods.[243]
Thou, Bolt of Indra, Vanguard of the Maruts, close knit to Varuṇa and child of Mitra – As such, accepting gifts which here we offer, receive, O Godlike chariot, these oblations.[244]

It is notable that even the chariot-horses were endowed with divine attributes:

O horse! Yoked to the chariot you became fast moving like wind.
By your dexterity you augment the glory of Indra. May repository of knowledge, the Maruts, yoke you to the chariot and Tvaṣṭā put agility into your legs.[245]

Due to the divine nature of the chariot, hymns were composed to pay obeisance to those who fought mounted on chariots and those who made the chariots.[246] Prayers were addressed to Bṛhaspati to conquer evil-doers and protect the chariots.[247]

In the *Atharvaveda*, various components of divine chariot are equated with the universal phenomenon. The four mares, which pulled the chariot represented the four directions, hoofs the *puroḍāśa* (offering), the upper-part the *antarikṣa* (sky), the two sides the earth and the heaven, the ropes the seasons, the mid-part the *kiṅkara* (slave), the voice the *parirathyam* (road), *Saṁvastara* (era) the chariot, *parivastara* (year) the seat, *virāḍa* (universe) the yoke, Agni the face, Indra the *rathī* (master of chariot) and Candramā the *sārathi* (charioteer).[248]

CHARIOT A GIFT ITEM

The chariots were often given as gift to the priests, perhaps for use as transport:

His who hath many gifts to give the Pajras (a *ṛṣi*), a chief who makes me rich in cars and horses.[249]
Horses of dusky colour stood beside me, ten chariots Svanaya's gifts, with mares to draw them.[250]
Ten cars with extra steed to each. . . . Hath Aśvatha to Pāyu (a *ṛṣi*) given.[251]
(Priest got a gift of) two chariots from Sudās with mares to draw them.[252]

CHARIOTS OF GODS

Agni

There are many hymns in the *Ṛgveda*, which refer to the chariot of Agni:

Thou farest as an envoy, having harnessed, Sublime One! Thy strong muscled radiant stallions (*Ṛv* 4.2.2).

Yoking red horses to and fro thou goest between you Deities and mortal races. (*Ṛv* 4.2.3).

This envoy joyeth in all seats of worship, Borne on his golden car (*Ṛv* 4.1.8).

The golden car of Agni was thus pulled by two strong muscled and radiant red horses.

Indra

The chariot of Indra named Nārāśaṁsa was created by Tvaṣṭā, the divine carpenter. Many hymns of the *Ṛgveda* refer to it:

Whose pair of tawny horses yoked in battles foemen challenge not (*Ṛv* 1.5.4).
On both sides to the car they yoke the two bay coursers dear to him (*Ṛv* 1.6.2).
Sixfold they bear him, or by fives are harnessed (*Ṛv* 3.55.18).
Here Tvaṣṭṛ, yoking to the car the Bay steeds (*Ṛv* 6.47.19).
For him the god, who is invoked by many, the two swift bay steeds to the pole I harness (*Ṛv* 3.35.2).

Let thy Bay stallions, harnessed, bring thee hither, Steeds with strong chariot and strong reins to hold them. (*Ṛv* 6.44.19).

In the *Bṛhat Saṁhitā*, a later work, the chariot of Indra is described as a eight-wheeled vehicle, decked with jems and bright like autumn sun.[253] The chariot of Indra was thus pulled by two yellowish brown horses. But in some cases more than two horses were yoked. Indra was generally accompanied by a charioteer though occasionally he himself managed the horses.

MARUTS

Chariot of Maruts, the Divine warriors, is widely referred to in the *Ṛgveda:*

And with their chariot tires they cleave the rock asunder in their might (*Ṛv* 5.52.9).
When to your car-poles, ye have yoked your spotted deer to be your steeds (*Ṛv* 5.55.6 also 1.85.5).
Bind to your car the bright red mares, yoke the red coursers to your car (*Ṛv* 5.56.6).
Good horses and good cars have ye (*Ṛv* 5.57.2).
Your weapons in your cars (*Ṛv* 5.57.6).
Your axle turns together both the chariot wheels (*Ṛv* 1.166.9).
Golden and dustless were their cars invested with their great strength (*Ṛv* 6.66.2).

The chariot of Maruts was thus, pulled by bright red mares of a superior breed. The wheels of their chariot were sharp edged and left a mark on the road they passed through. The war chariot of Maruts is said to have sharp scythe like blades attached to the wheel. (*Ṛv* 1.166.10). The weapons were also stacked in the chariot. Yoking of spotted dear to their chariot is difficult to explain.

SŪRYA (SUN)

The following hymns refer to the chariot of the Sun:

Seven Bay steeds harnessed to thy car bear thee (*Ṛv* 1.50.8).
Seven to the one-wheeled chariot yoke the courser; bearing seven names the single courser draws it.

> Three-naved the wheel is, sound and undecaying (*Ṛv* 1.164.2).

It appears that the chariot of Sun was pulled by seven horses. It was mounted on one wheel, which was three naved.

Aśvins

The chariot of Aśvins is described in the following hymns:

> Come, O ye Aśvins, mounted on your triple car three seated, beautiful of form (*Ṛv* 1.47.2).
> Come to us with your chariot triple seated, three-wheeled, of triple form, that rolleth lightly (*Ṛv* 1.118.2;1.34.2; 1.47.2).
> O Aśvins, let your falcons bear you hither, yoked to your chariot, swift with flying pinions (*Ṛv* 1.118.4).
> May your winged coursers, best to draw, Nāsatyas! Convey you to the object of your wishes (*Ṛv* 6.63.7).
> May your gold chariot, drawn by vigorous horses, come to us (*Ṛv* 7.69.1; also 4.44.4, 8.5.35).

It appears that the chariot of Aśvins was a three-wheeled vehicle, providing for three seats, two for Aśvins and one for charioteer. The horses have been compared with falcon and described as winged because they ran very fast.

Uṣas (Dawn)

Like other Vedic divinities, Uṣas also travelled in a chariot as borne out by the following hymns:

> This dawn hath yoked her steeds afar, Borne on a hundred chariots she, auspicious dawn advances (*Ṛv* 1.49.7).
> All this with red-rayed steeds have they divided: the Dawn on bright cars shine in wondrous fashion (*Ṛv* 6.65.2).
> Let docile horses of far-reaching splendour convey thee hitherward, the golden coloured (*Ṛv* 3.61.2).

The golden hued chariot of Uṣas was resplendent and splendorous. It was pulled by red horses.

Vāyu

The chariot of Vāyu is described in the following hymns:

O Vāyu, on refulgent car come to the drinking of the juice.
Harness, O Vāyu, to thy car a hundred well-fed tawny steeds, yea, or a thousand steeds, and let thy chariot come to us with might (*Ṛv* 4.48.4-5).
Two red steeds Vāyu yokes, Vāyu two purple steeds, swift-footed, to the chariot, to the pole to draw, most able, at the pole, to draw (*Ṛv* 1.134.3).
O Vāyu, come to us with all the thousand chariots that are thine (*Ṛv* 2.41.1).

The chariot of Vāyu was pulled by two strong tawny steeds. Harnessing of one hundred or one thousand chariot is indeed wild imagination.

Mitra-Varuṇa

The chariot of Mitra-Varuṇa is also mentioned in some hymns:

Let your well-harnessed horses bear you hither: hitherward let them come with reins drawn tightly (*Ṛv* 5.62.4).
Ye mount your car gold-hued at break of morning (*Ṛv* 5.62.8).

The car of Mitra-Varuṇa was gold-hued. It was managed with reins drawn tightly.

Savitar

Savitar's chariot is referred to in the following hymns:

Borne in his golden chariot he cometh (*Ṛv* 1.35.2).
With two bright Bays, adorable, he journeys (*Ṛv* 1.35.3).
His chariot decked with pearl, of various colours, lofty, with golden pole, the God hath mounted (*Ṛv* 1.35.4).
Drawing the gold-yoked car his Bays, white footed, have manifested light to all the peoples (*Ṛv* 1.35.5).

The chariot of Savitar was golden in hue and was pulled by white footed reddish brown horses.

It may be mentioned here that all the Vedic gods moved in chariots pulled by one or more horses. Perhaps, some kind of symbolism was associated with the number of horses which is not always understandable. Gods like Agni, Vāyu, Aśvins, Indra, etc.

joined the Soma ritual mounting their chariots. But the chariots were more widely used in battles against the Asuras and Dasyus by Indra and his devotees.

ARCHAEOLOGICAL EVIDENCE

Here it may be mentioned that no visual representation of chariot, which could be associated with the Vedic Āryans with certainty, has yet been discovered. But a closer look at a few chariot carvings in some rock-shelters opens up some new possibilities. As the primitives living in these rock-shelters could not have invented a chariot or say even a cart the subject of carving might have been borrowed by them from outside. If these carvings could be dated around 1000 BC, it was perhaps the Vedic chariot which the primitive artist was trying to portray.

In this connection two carvings, one from Wyndham in Mirzapur (Figure 2.1) and the other from Chibba Nala in Chambal Valley, depicting a vehicle are of unusual interest (Figure 2.2). Jagdish Gupta is inclined to indentify Wyndham Vehicle as a cart, though he is unsure about the identity of the animals yoked to it.[254] But these animals could not have been anything but a horse is evident from the following facts: The hair on the neck and tail clearly show it to be a horse and the absence of horns and hump indicate that it was not an ox. The body-length of the animal and the big forward-leap also confirm that it is a horse.

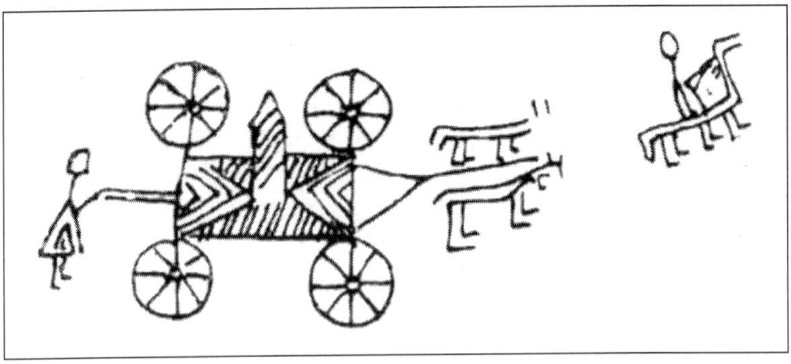

Figure 2.1: Battle Scene – Chibba Nala rock-shelter (Chambal Valley)

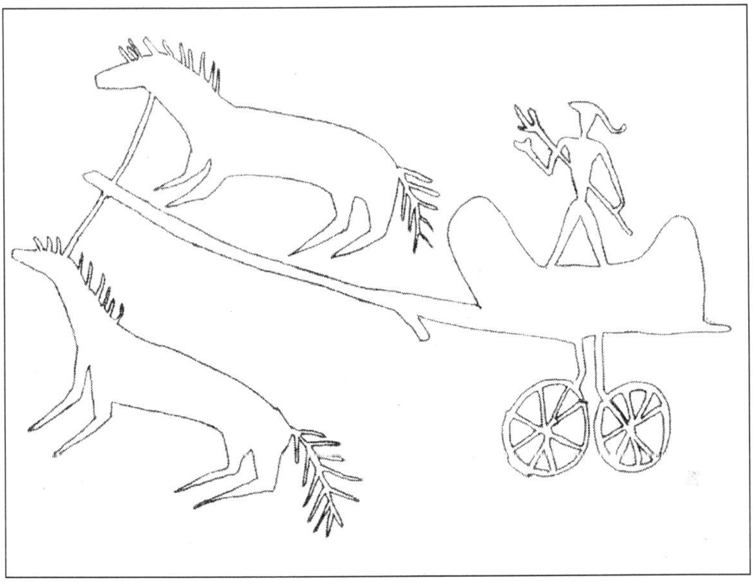

Figure 2.2: Chariot with Mounted Warrior, Wyndham (Mirzapur) Rock-shalter

Further, the person mounting the vehicle is holding a weapon like trident in his hand. It is likely that, the carving portrays a warrior going to war or on a hunting spree. The vehicle could be identified as a chariot also because warriors do not go for battle or for hunting, mounting a cart.

Another carving from Chibba Nala rock-shelter also adduces evidence on chariot. Here, a two-wheeled vehicle is being pulled by two horses. A three-pronged pole connects the wheels to the yoke. It was truly a chariot is evident from the fact that the carving is a battle-scene. A warrior is depicted as running away from the battlefield. Trapped by the enemy, equipped with bow and arrow, he surrenders with raised hands. Gupta, however, identifies this vehicle as cart.[255] But the wheels depicted in the carvings are spoked suggesting an advanced stage of chariot building.

Incidentally, all wheels depicted in the carvings are spoked.

In this connection, a reference could also be made to two cave illustrations of chariots at Morhana Pahar in Mirzapur district.

According to Allchin, one of these chariots being attacked by two men on foot with bow and arrow and shield and spear record the reminiscence of a sortie in the early centuries BC from some centre of Gaṅgā-Yamunā Doab into the territory of hunting tribes. One of these chariots is being drawn by two and the other by four horses.[256]

NOTES

1. *आ द्वाभ्यां हरिभ्यामिन्द्र या ह्या चतुर्भिरा षड्भिर्हूयमानः।*
   अष्टाभिर्दशभिः सोमपेय मयं सुतः सुमख मा मृधस्कः।।
   आ विंशत्या त्रिंशत्या याह्यर्वाङा चत्वारिंशता हरिभिर्युजानः।
   आ पञ्चाशता सुरथेभिरिन्द्रा आ षष्ट्या सप्तत्या सोमपेयम्।।
   आशीत्या नवत्या याह्यर्वाङा शतेन हरिभिरुह्यमानः।
   Ṛgveda, 2.18.4
   The number of horses is highly exaggerated as not more than four horses could have been yoked to the chariot. According to A.L. Basham 'the Vedic chariot was a light chariot with spoked wheels, drawn by two horses yoked abreast, and carrying two warriors'. *The Wonder that was India*, p. 37; according to Macdonell 'Vedic Āryans fought either on foot or on chariot. The latter had two occupants, the fighter and the driver. This was still the case in the *Mahābhārata*, where we find Kṛṣṇa acting as charioteer of Arjuna', *A History of Sanskrit Literature*, p. 165.
2. A. Cotterell, *Chariot*, p.36; according to David Frawley the migration took place in the reverse direction. 'Parhaps the Āryan peoples of Central Asia and Europe migrated from India via Gandhāra', *Gods, Sages and Kings*, p. 83.
3. Zonaide A. Ragozin, *History of Vedic India*, p. 49.
4. A. Parpola, 'The Nāsatyas, the Chariot and Proto-Āryan Religion', paper read at 50th International Conference of Eastern Studies, in 2005 at Kyoto, p. 4.
5. Ghirshman quoted by Parpola, ibid., p. 5.
6. Adolf Kaegi, *The Life in Ancient India*, p. 19.
7. *एतं ते स्तोमं तुविजात विप्रो रथं न धीरः स्वपा अतक्षम्।*
   Ṛgveda 5.2.11; also 1.61.4, 1.130.6, 1.141.8
8. *तक्षन नासत्याभ्यां परिज्मानं सुखं रथम्।* Ibid., 1.20.3
9. *शच्या हरी धनुतरावतष्टेन्द्रवाहावृभवो वाजरत्नाः।* Ibid., 4.35.5; also 4.33.8.

10. *रथं ये चक्रुः सुवृतं सुचेतसोऽविह्वरन्तं मनसस्परिध्यया।* Ibid., 4.36.2; also 33.8.
11. *Ṛgveda*, 10.29.12.
12. *अनवस्ते रथमश्वाय तक्षन्।* Ibid., 5.31.4; Griffith associates Anus with Bhṛgus. *The Hymns of the Ṛgveda*, p.250, fn. 4.
13. *तक्षाम भृगवो न रथम्।* 10.39.14.
14. *इन्द्रस्य ऋभुक्षा वरुणस्य विभ्वा।* Ibid., 4.33.9; It is said that the chariot built by Vibhvā excelled in war. Ibid., 4.36.5.
15. Ibid., 4.33.9
16. *Yajurveda*, 21.55; In the *Ṛgveda* (1.32.2, etc.) he is mentioned as the maker of Indra's *vajra*.
17. *रथो न यातः शिक्कभिः कृतो।* *Ṛgveda*, 1.141.8.
18. D.D. Kosambi, *The Culture and Civilizations of Ancient India in Historical Outline*, p. 80.
19. *Ṛgveda*, 4.34.1.
20. *ये धीवानो रथकाराः कर्मारा ये मनीषिणः।* *Atharvaveda*, 3.5.6.
21. *Taittirīya Brāhmaṇa*, 1.1.4.8; *Āpastamba Śrautasūtra*, 6.1.50.
22. *Pūrva-Mīmāṁsā Sūtra*, 6.1.50; For more details on the social status of *rathakāra* see P.V. Kane, *History of Dharmaśāstra*, vol. II, part I, pp. 45-6; also F. Staal, *Discovering the Vedas*, pp. 62-8.
23. *Taittirīya Brāhmaṇa*, 1.1.4.
24. *Baudhāyana Gṛhyasūtra*, 2.4.6.
25. *रथकारेभ्यश्च वो नमः।* *Yajurveda*, 16.27.
26. *कक्षीवन्तं यदी पुना रथं न कृणुथो नवम्।* *Ṛgveda*, 10.143.1.
27. *रथं न दस्रा करणा समिन्वथः।* Ibid., 1.119.7.
28. *Śatapatha Brāhmaṇa*, 12.5.1.5.
29. S.D. Singh, *Ancient Indian Warfare with Special Reference to Vedic Period*, p. 65.
30. Ibid.
31. J. Keegan, *A History of Warfare*, p. 167.
32. F. Staal, op. cit., p. 17.
33. Ibid., p. 36.
34. *रथो वां मित्रवरुणा दीर्घाप्साः स्यूमगभस्तिः।* Ibid., 1.122.15.
35. *Vedic Index*, vol. II, p. 202; also see pp. 201, 203
36. *Āpastamba Śulbasūtra*, 2.6.15-19.
37. *Baudhāyana Śulbasūtra*, 1.10-12.
38. *Katyayana Śulbasutra*, 2.1-4.
39. *अष्टाशीतिः शतमीषा तिर्यगक्षश्चतः शतम्।*
षडशीतिर्युगं चास्य रथचारक्य उच्यते।।
Manava Śrautasutra, 10.1.2.1.

40. For details see chapter 'Brāhmaṇas and Sūtras'.
41. रथमात्री निरुढपशुवन्धस्य वेदिर्भवतीति विज्ञायते। *Āpastamba Śulbasūtra*, 2.6.7.
42. तस्य खल्वाहु रथाक्षमात्री पश्चात्तिर्यग्।
ईषया प्राची। विपथयुगेन पुरस्तात। Ibid., 2.6.8-10. *Vipatha* in stated to be a kind of chariot, perhaps the same as *cāraṇa*, but different from *krīḍāratha*, a luxury car.
43. अष्टाशीतिशतमीषा। Ibid., 2.6.15.
44. तिर्यगक्षश्चतुश्शतम्। Ibid., 2.6.16.
45. षडशीतियुगं च। Ibid., 2.6.17.
46. S. Piggott, *Prehistoric India*, pp. 281-3.
47. M. Sparreboom, *Chariots in the* Veda, p. 25.
48. अग्ने सुखतमे रथे देवां इलित आ वह। *Ṛgveda*, 1.13.3.
49. इन्द्रं सुखतमे रथे। Ibid., 1.16.2; also 1.20.3.
50. Ibid., 1.35.4-5.
51. हिरण्यरथं। Ibid., 1.30.16.
52. हिरण्ययेन सविता रथेना। Ibid., 1.35.2; Chariots wrought with gold were also known in contemporary Western Asia. Nic Fields, op. cit., p. 42.
53. युक्ष्वा ह्यरुषी रथे हरितो देव रोहितः। Ibid., 1.14.12; also 1.94.10, 1.45.2, 8.63.14.
54. व्याघ्रो अधि वैयाघ्रे विक्रमस्व दिशो महीः। *Atharvaveda*, 4.8.4.
55. *Ṛgveda*, 6.75.8; Griffith explains *rathavāhanam* as a platform, stand or truck on which the chariot was placed when not in use. op. cit., p. 131. He says that the word *rathavāhanam* seems in this place to mean also the oblation offered by the warrior to the ideal war chariot personified, or to a tutelary deity of chariots. *The Hymns of the Ṛgveda*, p. 331.
56. *Ṛgveda*, 3.53.6.
57. रथवाहणं हविरस्य नाम यत्रायुधं निहितमस्य वर्म। *Yajurveda*, 29.45.
58. *Ṛgveda*, 6.47.26-27.
59. अभि व्ययस्व खदिरस्य सारमोजो धेहि स्पन्दने शिंशुपायाम्। Ibid., 3.53.19.
60. सुकिंशुकं शल्मलिं विश्वरूपं हिरण्यवर्णं सुवृतं सुचक्रम्। Ibid., 10.85.20.
61. वनस्पते वीड्वङ्गो हि भूया अस्मत्सखा प्रतरणः सुवीरः
गोभिः संनद्धो असि वीडयस्वाआस्थाता ते जयतु जेत्वानि।
दिविस्पृथिव्याः पर्योज उद्भृतं वनस्पतिभ्यः पर्याभृतं सहः
अपामोज्मानं परि गोभिरावृतमिन्द्रस्य वज्रं हविषा रथं यज।
Ātharvaveda, 6.125.1-2; also see *Ṛgveda*, 6.47.26-27.
62. *Ṛgveda*, 6.47.27.
63. इन्द्रस्य वज्रो मरुतामनीकं मित्रस्य गर्भो वरुणस्य नाभिः
सेमां नो हव्यदातिं जुषाणो देव रथ प्रति हव्या गृभाय। Ibid., 6.47.28.

64. T.H. Griffith Ralph, *The Hymns of the Ṛgveda*, p. 331.
65. A. Cotterell, *Chariot*, p. 93.
66. F. Staal, op. cit., pp. 16-17.
67. Ibid., p. 17.
68. *ग्रीवायां वद्धो अपिकक्ष आसनि।* *Ṛgveda*, 4.40.4.
69. *उत ते सुष्टुता हरी वृषणा वहतो रथम्।* Ibid., 8.13.23; also 5.30.1, 5.36.5, 6.23.1.
70. *इन्द्रमित् केशिना हरी सोमपेयाय वक्षतः।* Ibid., 8.14.12.
71. Ibid., 8.1.25.
72. *युञ्जन्त्यस्य काम्या हरी विपक्षसा रथे।* Ibid., 1.6.2; also 1.82.4-6, 1.5.4
73. *युक्ष्वा हि केशिना हरी।* Ibid., 1.10.3.
74. Ibid., 10.22.4; 1.134.3.
75. Ibid., 8.63.14, 1.94.10.
76. *सप्त त्वा हरितो रथे वहन्ति देव सूर्य।* Ibid., 1.50.8.
77. *चत्वारिंशद दशरथस्य शोणाः सहस्रस्याग्रे श्रेणि नयन्ति।* Ibid., 1.126.4.
78. Ibid., 4.48.4-5; also 6.47.18.
79. *आ त्वा सहस्रमा शतं युक्ता रथे हिरण्यये।* Ibid., 8.1.24.
80. Ibid., 2.18.4-6. In the *Bṛhat Saṁhitā* of Varāhmihira (43.6) a late work the chariot of Indra is said to be eight wheeled.
    तं विष्णुतेजोद्मवअष्टचक्रे रथे स्थितं भास्वति रत्नचित्रे।
    देवीप्यमानं शरदीय सूर्यं ध्वजं समासाद्य सुमोद शक्रः।।
81. Ibid., 10.101.11; also see Wilson, op. cit., vol. 4, p. 51; In ancient Rome a two-horsed chariot was called *biga*, a three-horsed chariot *triga* and a four-horsed chariot *quadriga.*
82. Ibid., 10.33.5
83. *युक्ष्वा ह्यरुषी रथे हरितो देव रोहितः। ताभिर्देवां इहा वह।* Ibid., 1.14.12.
84. Ibid., 5.56.6.
85. Ibid., 8.13.23, 8.1.24.
86. *तद रासभो नासत्या सहस्रमाजा यमस्य प्रधने जिगाय।* Ibid., 1.116.2; also 1.34.9.
87. *Chāndogya Up.*, 4.2.1.
88. *Vādhūla Śrautasūtra* quoted by Ram Gopal, p. 515; Also *Kātyāyana Śrautasūtra*, 22.10.31.
89. Ibid., 1.85.4.
90. Ibid., 10.102.4; 3.53.17-18
91. *युक्ष्वा हरुषी रथे।* Ibid., 1.14.12.
92. *सप्त त्वां हरितो रथे वहन्ति देव सूर्य।* Ibid., 1.50.8.
93. *यदयुक्था अरुषा रोहिता रथ।* Ibid., 1.94.10; also 1.45.2.
94. Ibid., 7.75.6.

95. Ibid., 2.17.3, 4.35.5; At another place (1.6.2) these are called red (*soṇā*). But in later literature these are described as green.
96. Ibid., 7.97.6.
97. *वायुर्युङ्क्ते रोहिता वायुररुणा।* Ibid., 1.134.3.
98. Ibid., 4.32.17. Indra had 1,000 trained horses.
99. Ibid., 2.18.1; According to Griffith, the rich new car symbolizes the morning sacrifice, four yokes the four pair of stones for pressing out the *soma* juice, the three whips the three tones of prayer and the seven reins the seven metres, op. cit., pp. 141-2.
100. *या वां कशा मधुमत्यश्विना सूनृतावती।* Ibid., 1.22.3
101. Ibid., 6.75.13; The term *अश्वाजनि* denotes whip; also *Yajurveda*, 29.50
102. *यत्र मन्थां विबध्नते रश्मीन् यमितवा इव।* Ibid., 1.28.4; also 4.36.1
103. Ibid., 1.38.12
104. *संशितो रश्मिना रथः संशितो रश्मिना हयः।* *Yajurveda*, 23.15.
105. R.L. Mitra, *Indo-Āryans*, p. 388; For details of Assyrian harness see Layard, *Nineveh*, vol. II, p. 354.
106. R.L. Mitra, op. cit., p. 388.
107. *Ṛgveda*, 10.85.20; *kiṁśuka* was the wood of the *Butea frondosa* while *śālmali* was *Salmalia malabarica*, i.e. silk cotton tree. The mention of two woods suggests that one was used for making the rims while the other, the harder one, was used to make spokes.
108. *Ṛgveda*, 2.39.3.
109. *अक्षो वश्चक्रा समया वि वावृते।* Ibid., 1.166.9; also 8.5.29.
110. *त्रिचक्रेण।* Ibid., 1.118.2; also 1.34.2, 9, 1.157.3, 10.41.1.
111. *सप्तचक्रं सप्त वहन्त्यश्वाः।* Ibid., 1.164.3, 1.34.9.
112. Ibid., 2.40.3.
113. Ibid., 1.164.2
114. *अराणां न चरमः।* Ibid., 8.20.14.
115. *अरान्न नेमिः परिभूरजायथाः।* Ibid., 1.141.9; also 6.54.3.
116. Ibid., 10.78.4; Some burial sites of Sintashta-Petrovka culture in modern Russia and Kazakhstan, dated around 2000 BC, have also yielded spoke-wheeled chariots.
117. In the *Ṛgveda*, five- and six-spoked wheels are referred to. Ibid., 1.164.12-13. Singh says that Mycenaean, Egyptian and early Hittite chariot wheel had four spokes, late Hittite and Assyrian wheel had six to eight spokes and Homeric wheel had eight spokes. He adds that the Vedic chariot may have had four to

eight spokes. S.D. Singh, op. cit., p. 27; also see R.L. Mitra, *Indo-Āryans*, p. 342.

118. The wheel of time (*cakram ṛtasya*) with its 12 spokes turns around and around (in) the sky and never ages. *Ṛgveda*, 1.164.11.
119. S.D. Singh, op. cit., p.27
120. *विभुरराज नेमिः परिभूरजायथाः* *Ṛgveda*, 1.141.9.
121. *अक्षो वश्चक्रा समया वि वावृते।* Ibid., 1.166.9.
122. T.H. Griffiths Ralph, op. cit., p. 115.
123. U.P. Thapliyal, *Warfare in Ancient India*, p. 481; There is also a view that Cyrus the Great, the Persian ruler had invented the scythe. In the Persian scythed chariot, the blade extended horizontally for a meter on the sides of the chariot. These chariots were used in the battle of Cunaxa and Gaugamela. Early Greeks and Romans also used such chariots.
124. *Ṛgveda*, 1.88.2.
125. Ibid., 1.38.12.
126. S.D. Singh, op. cit., p.27.
127. S. Piggott, *Prehistoric India*, p. 278.
128. *गोभिः संनद्धो असि।* *Atharvaveda*, 6.125.1-2.
129. *Ṛgveda*, 8.46.27.
130. *अक्षो हिरण्ययः।* Ibid., 8.5.29.
131. Ibid., 3.53.19.
132. *Ṛgveda*, 2.12.8, 3.14.3.
133. Ibid., 1.34.9, 1.47.2, 1.118.1-2.
134. *अष्टाबन्धुरम्।* Ibid., 10.53.7.
135. *इन्द्रः सव्यष्ठाश्चन्द्रमाः सारथिः।* *Atharvaveda*, 8.8.23.
136. *Ṛgveda*, 3.6.6.
137. Ibid., 3.53.17.
138. Ibid., 1.5.4.
139. Ibid., 1.6.2.
140. Ibid., 1.82.4.
141. Ibid., 1.84.6.
142. In the Vedas and the epics the charioteer is variously named as Adhyāsthātṛ, Abhīśugraha, Niyantṛ, Parṣṇiyantṛ, Prajitṛ, Yantṛ, Rathakuṭumbin, Rathagṛtsa, Rathayantāru, Rathin, Samgrahītṛ, Savyaṣṭhā, Sārathi, Sūta, Sthatṛ. M. Sparreboom, *Chariots in the Veda*, p.134.
143. Ibid., 5.83.3; 6.75.6; also *Vedic Index*, p. 202; the term *aśvājani* has also been used for whip.
144. *Ṛgveda*, 1.82.6

145. नकिष्ट्वद् रथीतरो हरी यदिन्द्र यच्छसे। *Ṛgveda*, 1.84.6.
146. तिष्ठा रथमधि तं वज्रहस्ताआरश्मिं देव यमसे स्वश्व: Ibid., 5.33.3.
147. Ibid., 8.45.7.
148. तिष्ठा रथमधि यं वज्रहस्ता रश्मीन् देव यमसे स्वश्वान्। *Yajurveda*, 10.22.
149. *Ṛgveda*, 6.75.6.
150. यमिष्ठास: सारथयो य इन्द्र ते। *Ṛgveda*, 1.55.7; It seems that occasionally more than one charioteer were engaged.
151. Ibid., 6.20.5; also 2.19.6, 5.31.9.
152. Ibid., 6.29.2.
153. Ibid., 2.12.8; The two included the warrior and the charioteer.
154. Ibid., 6.75.6.
155. अभिशूरिव सारथि:। Ibid., 6.57.6.
156. Ibid., 6.56.2.
157. Ibid., 6.55.2, 6.56.3.
158. Ibid., 6.57.6.
159. Ibid., 2.19.6.
160. Ibid., 5.31.9.
161. Ibid., 6.20.5.
162. इन्द्र: सव्यष्टा चन्द्रमा सारथि:। *Atharvaveda*, 8.8.23.
163. V.R.R. Dikshitar, *War in Ancient India*, p. 158.
164. यं रथं अधितिष्ट स्वश्वान रश्मीन आयससे।
तिष्ठा रथमधि यं वज्रहस्ता रश्मीन देव यमसे स्वश्वान। *Yajurveda*, 10.22.
165. *Ṛgveda,* 6.75.13.
166. रथे तिष्ठन नयति वाजिन: पुरो यत्र-यत्र कामयते सुषारथि:। *Yajurveda*, 29.43
167. अश्वाजनि प्रचेतसोअश्वान्त्समत्सु चोदय। Ibid., 29.50
168. *Ṛgveda*, 6.75.6.
169. *Atharvaveda*, 15.2.7.
170. *Ṛgveda*, 4.34.1.
171. *Vedic Index*, vol. II, p. 200; *Śukla Yajurveda*, 30.6-7.
172. जीमूतस्येव भवति प्रतीकं यद् वर्मी याति समदामुपस्थे।
अनाविद्धया तन्वा जय त्वं स त्वा वर्मणो महिमा पिपर्तु। *Ṛgveda*, 6.75.1.
173. मर्माणि ते वर्मणा छादयामि सोमस्त्वा राजामृतेनानु वस्ताम्। Ibid., 6.75.18.
174. अहिरिव भोगै: पर्येति बाहुं ज्याया हेतिं परिबाधमान:।
हस्तघ्नो विश्वा वयुनानि विद्वान् पुमान् पुमांसं परि पातु विश्वत:।। Ibid., 6.75.14. *Godhā* was also worn to protect the hand. Ibid., 8.69.9.
175. शिप्रा शीर्षसु वितता हिरण्ययी:। Ibid., 5.54.11.
176. ये अञ्जिषु ये वाशीषु स्वभानव: स्रक्षु रुक्मेषु खादिषु।
श्राया रथेषु धन्वसु। Ibid., 5.53.4.

177. अंसेषु व ऋष्टयः पत्सु खादयो वक्षःसु रुक्मा मरुथो रथे शुभः। Ibid., 5.54.11.
178. वाशीमन्त ऋष्टिमन्तो मनीषिणः सुधन्वान इषुमन्तो निषङ्गिणः। स्वश्वाः स्थ सुरथाः। Ibid., 5.57.2.
179. Ibid., 6.7.52; 5.53.4.
180. Ibid., 6.75.11.
181. Ralph T.H. Griffith, op. cit., p. 331 fn.
182. आलाक्ता। *Ṛgveda*, 6.75.15.
183. इषुधिः सङ्काः पृतनाश्च सर्वाः पृष्ठे निनद्धो जयति प्रसूतः। Ibid., 6.75.5.
184. अश्वाजनि प्रचेतसो अश्वान त्समत्सु चोदय। Ibid., 6.75.13.
185. इन्द्रः सव्यष्ठाश्चन्द्रमाः सारथिः। *Atharvaveda*, 8.8.23; In *Taittirīya Saṁhitā* सव्येष्ठ-सारथी combine means the warrior and the charioteer (1.7.9.1).
186. *Ṛgveda*, 1.34.9; 1.47.2; 1.118.1.
187. S. Piggott, *Prehistoric India*, p. 281.
188. *Ṛgveda.*, 6.75.9.
189. अस्माकमिन्द्र रथिनो जयन्तु। *Yajurveda*, 29.57.
190. *Ṛgveda*, 4.38.6.
191. रथीतमो रथीनाम्। Ibid., 8.45.7; also *Yajurveda*, 17.61.
192. वित्वक्षणः समृतौ चक्रमासजो। Ibid., 5.34.6.
193. Ibid., 5.35.5.
194. Ibid., 5.35.7.
195. Ibid., 10.102.11.
196. Ibid., 6.26.4.
197. Ibid., 6.31.5.
198. Ibid., 6.18.9.
199. Ibid., 5.57.2.
200. Ibid., 9.83.5.
201. वृहस्पते परि दीया रथेन रक्षोहाऽमित्रां अपबाधमानः। प्रभञ्जनत्सेनाः प्रमृणो युधा जयन्नस्माकमेध्यविता रथानाम्। *Yajurveda*, 17.36.
202. ये रथिनो ये अरथा असादा ये च सादिनः। सर्वानदन्तु तान् हतान् गृधाः श्येनाः पतत्रिणः।। *Atharvaveda*, 11.10.24.
203. आ तेन यातं मनसो जवीयसा रथम्। Ibid., 10.39.12.
204. अभ्रियेव वातः। *Ṛgveda*, 1.116.1.
205. Ibid., 6.46.13-14.
206. Ibid., 8.47.5.
207. त्रीणि शतानि राजपुत्राणामुभयालोहकवचानां सन्नधसारथीनां विततवरूथानां चतुर्युजां; त्रीणि शतान्यराज्ञां सन्नधानासन्नधसारथीनां विततवरूथानां त्रियोगाणां; त्रीणि शतानि सतग्रामणीनां-सन्नाधानामसन्नधसारथीनां विततवरूथानां द्वियौगानां; त्रीणि शतानि

क्षत्तृसंग्रहीतृहणां विपथपत्तीनां. . . अथास्तंयत्यादित्य एते राजपुत्राश्चतुर्धा व्युद्यन्ति पञ्चसप्तति: पञ्चसप्तति: प्रतिदिशम्। *Vādhūla Śrautasūtra*, 11.10.1.

208. Nic Fields, op. cit., p. 17.
209. R.L. Mitra, *Indo-Āryans*, p. 340.
210. Ibid., pp. 340-1.
211. S. Piggott, op. cit., pp. 268-9.
212. Ibid., p. 253, The close affinity between the Mitannians and Indo-Āryans is also confirmed by the fact that the Mitannian Mattiuaza, son of Dusratta, in a treaty concluded by him with Hittite king in about 1380 invokes his gods as witnesses. These gods have been identified with Vedic gods Mitra, Varuṇa, Indra and Nāsatyas (Aśvins). Ibid., p. 252.
213. A.C. Das, *Ṛgvedic Culture*, p. 226; Das adds that race-course was called *kāṣṭhā*, and that prizes (*dhana*) were offered. The institutor of race was called *āji-kṛt* or *āji-pati*. Ibid.
214. A. Kaegi, *The Ṛgveda*, p. 19.
215. *Ṛgveda*, 8.69.4-6; There are many more references to chariot-race. *Ṛgveda*, 1.116.2, 4.15.6, 6.53.4, 9.47.5, 9.53.2, 9.91.1, 10.64.6, 10.78.5.
216. Ibid, 6.45.12-15.
217. Ibid., 9.91.1.
218. Ibid., 9.53.2.
219. R.S. Sharma, *Aspects of Political Ideas and Institutions in Ancient India*, p. 165.
220. Ibid.
221. Ibid., p. 166; *अजिमेव अस्मिन् अजामहै स यो न उज्जेष्यति तस्य न इदम् भविष्यतीति।* *SB*, 5.1.3.3.
222. *Ṛgveda*, 1.51.4-6.
223. Ibid., 1.53.6-9.
224. Ibid., 4.16.12-13.
225. Ibid., 7.99.4-5.
226. *अयमग्नि: सत्पतिर्वृद्धवृष्णो रथीव पत्तीनजनत्पुरोहित:।*
नाभा पृथिव्यां निहितो दविद्युतदधस्पदं कृणुतां ते पृतन्यव:।। *Atharvaveda*, 7.64.1.
227. *इन्द्राग्नी काम सुरथं हि भूत्वा नीचै: सपत्नान् मम पादयथ:।* Ibid., 9.2.9.
228. Also *Ṛgveda*., 7.33; 7.83.
229. David Frawley, *Gods, Sages and Kings*, p. 104.
230. *Ṛgveda*, 1.10.3.
231. Ibid., 1.14.6.
232. Ibid., 1.14.12.

233. Ibid., 1.134.1, also 2.41.1.
234. Ibid., 4.14.1, also 4.14.4.
235. Ibid., 1.16.1-2.
236. Ibid., 1.22.2.
237. Ibid., 8.5.28-29.
238. Ibid., 1.35.2-5.
239. Ibid., 8.1.24.
240. Ibid., 8.1.25.
241. Ibid., 1.56.2.
242. Ibid., 6.47.26; also *Atharvaveda*, 6.125.1.
243. Ibid., 6.47.27; *Atharvaveda*, 6.125.2.
244. Ibid., 6.47.28; *Atharvaveda*, 6.125.3; The hymn also appears in the *Yajurveda* (29.54).
245. *नातरंहा भव वाजिन्युज्यमान् इन्द्रस्येव दक्षिणः श्रियौधि।*
युञ्जन्तु त्वा मरुतो विश्ववेदस आ ते त्वष्टा पत्सु जवं दधातु।।
Yajurveda, 9.8.
246. *नमो रशिभ्यो।* *Yajurveda*, 16.26.
रथकारेभ्यश्च वो नमो। Ibid., 16.27.
आशुरथाय च नमः। Ibid., 16.28.
247. *युधा जयन्नस्माकमेध्यविता रथानाम्।* Ibid., 17:36.
248. *दिशश्चतस्रोऽश्वतर्यो देवरथस्य पुरोडाशाः शफा अन्तरिक्षमुद्धिः।*
द्यावापृथिवी पक्षसी ऋतवोऽभीशवोऽन्तर्देशाः किंकरा वाक्परिरथ्यम्।।
संवत्सरो रथः परिवत्सरो रथोपस्थो विराडीषाग्नी रथमुखम्।
इन्द्र सव्यष्ठाश्चन्द्रमा सारथिः *Atharvaveda*, 8.8.22-23.
249. *Ṛgveda*, 1.122.8.
250. Ibid., 1.126.3.
251. Ibid., 6.47.24.
252. Ibid., 7.18.22.
253. *तं विष्णुतेजोद्भवअष्टचक्रे रथे स्थितं भास्वति रत्नचित्रै।*
देदीप्यमानं शरदीय सूर्यं ध्वजं समासाद्य मुमोद शक्रः।। *Bṛhat Saṁhitā*, 43.6.
254. *Vārṣikī'63*, Rajasthan Lalit Kala Academy, Jaipur, p.15; Gupta considers the rock-carving to be a production of Neolithic age.
255. Ibid., pp. 18-19.
256. B. Allchin, *Morhana Pahar: A Rediscovery*, *Man*, 1958, no. 58, pp. 153-5

CHAPTER 3

Brāhmaṇas and Sūtras

CHARIOT IN ESOTERIC REPRESENTATION

In the Brāhmaṇas and Sūtras, there are many references to chariots, particularly in relation to ceremonials like Vājapeya, Rājasūya and Aśvamedha sacrifices. It appears that by now, the chariot had evolved from a war machine and means of transport into an esoteric adorable object.

In the *Kaṭhopaniṣad*, the chariot is allegorized in an esoteric sense. Yama tells Naciketā that self is the chariot-rider, the body is the chariot, the intellect is the charioteer, the mind is the rein, the senses are the horses and the worldly objects the path over which these horses move about.[1] At another place, Yama makes an offer of a charming woman and a chariot to Naciketā, to desist him from asking a question.[2]

In the *Jaiminīya Brāhmaṇa*, the chariot is compared to the sacrifice.

one who knows the chariot of the gods becomes the possessor of chariots. The sacrifice is the chariot of the gods. The Bahiṣpavamāna is the front part of the chariot; Bṛhad and Rathantara are its two horses; the four Ājya stotras are its cords (yoktrāṇi); the two pavamānas are its reins (abhiṣu); the ausana and kāva sāmana are its two sides (pakṣasī); the naudhas and kālaya are its two axle pegs; the vāmadevya is its driver's seat; the *yajñāyajanīya* is its middle place; and the man (sacrificer) is one who sits in it.[3]

The *Aitareya Āraṇyaka* calls *prāṇa* as the chariot of gods.

Now there is a chariot of the god (*prāṇa*) destroying all desires. Its front part is speech, its wheels the ears, the horses its eyes and the driver the mind. *Prāṇa* mounts the chariot and by means of meditating on it, he reaches Hiraṇyagarbha. He who has no understanding and whose mind (the reins) is never firmly held, his senses (horses) are unmanageable like vicious horses of a charioteer.[4]

The frequent depiction of chariots in the Upaniṣads and Brāhmaṇas suggests that these were widely used for transport as well as in war.

In the *Lāṭyāyana Śrautasūtra*, various uses of the chariot including its make are described in detail.[5] In the making of a chariot, it is said, that stuff like leather, bone, iron, copper, gold, etc., were used and that it was a two-wheeled vehicle managed with right and left reins.[6] Horses, two,[7] three[8] or four[9] in number were yoked to the chariot depending on the need of the sacrificial rite. Mares, both he mules and she-mules, were also yoked to the chariot on certain occasions.[10]

VĀJAPEYA SACRIFICE

During the Vājapeya sacrifice, which was undoubtedly a military ritual, the king dressed up in armour. His followers also put on armour. The king did this at the command of the priest who told him to tie-up the protective armour, raise the weapons, and equip with three arrows made of iron, and a fourth one made of some other metal.[11] He, then, directed king to board the chariot.[12]

Meanwhile, preparations were made for the impending sham battle, in which the king was to emerge as a conqueror. Provision was made for the tracks of the chariot towards the south of the sacrificial altar. Two pieces of antelope-skin, one on the east and the other on the west, were provided to serve as targets, perhaps symbolizing the adversaries of the king. A number of cows were assembled beyond the target, i.e. the antelope-skin, to serve as booty for the victorious king.[13]

In the *Baudhāyana Śrautasūtra*, the narrative of a sham battle runs as follows:

Having ascended the chariot following the ritual, the *adhvaryu* (priest) makes the sacrificer recite seventeen *ujjiti* formulas (victory *mantras*) Then he gives him a whip for the horses with the verse, 'O strong whip, fast moving for winning food, do thou instigate the horses in battle'. He hits the horses with it in the order of their yoking: The southern one with 'thou art a steed', the middle one with 'thou art a horse', the northern one with 'thou art strong'. He (priest) instructs the participants of the chariot race, 'O horses do you run for food: do you run in the impulse of Maruts, do you traverse the distances, tread the way and reach the goal'. A pole of *Ficus glomerata* (*auḍumbara*) is fixed at a distance of seventeen throws of a yoke halter. It is called the goal. 'O participants, do not go beyond it, do not return without reaching it. Keeping it so as to be inside the wing (of the chariot), having controlled the left, lead by the right, do not start without being directed'.[14]

In the *Mānava Śrautasūtra*, the ritual is described with minor additions and deletions.

O steeds, win the race, establishing the way, measuring the leagues; attain the goal. . . . May the swift coursers, hearing the call, all hear our call, running strongly, winning a thousand, eager to gain in the gaining of praise, the steeds, that have won in the contests great prizes. This steed hastens at a gallop, bound at the neck, the shoulder and the mouth, acquiring strength like *dadhikrā* (divine horse) springing along the bends of paths. . . . There upon the king's son, who is nearest to him mounts. With: O swift whip, swift in the races, (make the steeds race in the contests) 'he takes the whip with: thou art a horse. Thou art a courser, thou art a runner' he throws it over the horses. Vaiśyas mount the other (chariots)., North of the pit they drive in a post of *uḍumbara* wood at a distance of seventeen bow-shots. He looks at it with: 'The goal'. Passing by it to the right they return.[15]

When the chariots have run back, he recites the verse: 'may the impulse of vigour come to me, may the heaven and earth with welfare to all, may the multiform fathers come to me, may Soma come to me with immortality'.[16] Some other rituals subsequently followed. Then followed a sham battle. The king in battle-dress mounting his chariot targetted the antelope-skin provided on the eastern side of the sacrificial altar. His followers, on the other hand, hit the antelope-skin provided on the western side. Once the king discovered the cows assembled beyond the target, he stopped shooting the arrows, thus in a way proclaiming his victory over the enemy and capturing his wealth symbolized by cows.[17]

On return, the chariot horses are unyoked after necessary rituals by Adhvaryu. The participants of chariot race then approached him and the latter gave a *kṛṣṇala* and a golden pot full of honey to each of them.[18]

This ritual of sham battle is also mentioned in the context of the Rājasūya sacrifice in the *Kaṇva Śatapatha Brāhmaṇa.*

On the northern side of the *Agnidhra*, he (*Adhvaryu*) places hundred or more than hundred of his cows. Having stationed them he brings down the chariot for him – whatever goes out of a warrior it is chased with a chariot; with a chariot whatever is to be won over is obtained.[19]

In the context of the Vājapeya sacrifice, there is an interesting reference to the chariot wheel. It is said that a post should be driven into the ground to the east of Āgnīdhriya shed. On top of that post a chariot wheel made of *uḍumbara* wood (*Ficus glomerata*) and having seventeen spokes should be placed. If a wheel of the desired specification was not available any other chariot wheel could be set to rotate on top of the post.[20] While the sacrificer and others ran the race mounting their chariots the priest chanted *mantras*, turning the wheel.[21] This ritual of turning the wheel seems to have granted the sacrificer (king) the status of a *cakravartin.*

Here it may be mentioned that in the *Baudhāyana Śulbasūtra*, two kinds of chariot wheels are mentioned, those with spokes (*sārāṇi*) and those with felly (*pradhis*).[22] In this context the spokes, nave and felloes of the wheel are also described in detail.[23]

CHARIOT-GIFT TO PRIESTS

The priests generally travelled to sacrificial venues mounting chariots.[24] They were often presented with chariots as sacrificial fee.[25] The priestly fee for Vājapeya sacrifice included seventeen chariots besides other things.[26] For two Apaciti sacrifices, the sacrifice performer was obliged to gift a chariot with four horses harnessed to it as fee.[27] The chariot was expected to be protected by bell-metal on both the sides[28] and the four horses yoked to it carried a value of 100 (perhaps *niṣkas*).[29] For Pañcavāttiya sacrifice, a chariot with two yoked horses and a third horse kept as stand by, and mounted by two charioteers was the fee (*dakṣiṇā*).[30] In the *Baudhāyana Śrauta-sūtra*, it is said that after a Rājasūya, the priests like Hotṛ, Udgātṛ and Adhvaryu were given lavish gifts, including a hundred chariots each. These chariots were strong enough to tread over an untrodden path.[31] The *Bhāradvāja Srautaśūtra* also refers to the gift of one fully furnished chariot to the priest.[32]

The *Lāṭyāyana Śrautasūtra* deals with the gifts in greater detail. It says that at the time of Mahāvrata sacrifice the sacrificer should present a chariot with horses along with 100 cows to the Brāhmaṇa priest.[33] The same present should be made to Adhvaryu, Udgātṛ and Hoṭr priests.[34] For a Trivṛt sacrifice, the priestly fee was a chariot drawn by four horses. The presents made to priest at the Vājapeya sacrifice comprised seventeen items, each seventeen in number, including chariots, elephants, horses, cows, etc.[35] The Udgātṛ was presented the horse, mounting which, the sacrificer ran the ceremonial race.[36]

The chariots presented to the priests as fee must have been used by them for travelling, as they were not supposed to be fighting a war. It is notable that priests alone did not travel in chariots. In the *Aśvalāyana Gṛhyasūtra* it is said that, 'if (the newly-married couple) have to make a journey (to their new home) let him cause her to mount the chariot with the (verse) may Pūṣan lead thee from here holding thy hand'.[37]

THE DIMENSIONS

The dimensions of the chariot, mentioned in the *Sulbasūtras*[38] have been discussed in the preceding chapter. On the basis of the evidence furnished by the Sūtras P.C. Chakravarti concludes:

> The masters of the Cartwright profession say that the *īśā* (the pole of the chariot which lies in the middle lengthwise – this has been compared with the centre line, the *pṛṣṭhya* or back-bone of a *vedī*) measures 188 (*aṅgulas*), perpendicular (*triyak*) to which *īśā* or *akṣa* (hind part of the car) equal to 104 (*aṅgulas*) and *yuga* (the forepart) equal to 86 (*aṅgulas*); such is the chariot (constructed according to the rules of the guild). These are the measures of the car.[39]

THE *RATHAKĀRA*

It is significant that by the time the Sūtras were composed the *rathakāras* (chariot-makers) as a sub-caste had slipped down in the social ladder. Now, the *rathakāras* do not appear participating in rituals like Vājapeya and Rājasūya, where chariots play a major role. According to the *Baudhāyana Dharmasūtra*, a son born out of the union of Śūdra female and Vaiśya male was called *rathakāra*.[40] In the *Āpastamba Śrautasūtra*, they are placed in a category separate from the three higher castes.[41] Still, a person belonging to *rathakāra* class could perform the rite called *ādhāna*.[42] Further, if necessary a high-born could also stay in his house during the night.[43]

MILITARY CONNECTION

In the *Baudhāyana Śrautasūtra* and *Lāṭyāyana Śrautasūtra*, there are many references to chariots in military context. Many words with prefix or suffix *ratha* are also mentioned. These include *rathagṛtsa* (skilled charioteer), *rathañjas* (brave in chariot fight), *rathasvana* (voice of a chariot), *rathacitra* (bright in chariot), *rathaprota* (attached to a chariot) and *asmaratha* (chariot of stone). The incidence of these words during the conduct of a sacrifice is indicative of military significance.[44]

The references to chariot in relation to Vājapeya sacrifice in *Baudhāyana* and *Lāṭyāyana Śrautasūtras* are similar excepting in some details. Baudhāyana says that, before the commencement of the race the chariot was placed on a wagon (*rathavāhanam*) towards the southern extension of the *mahāvedī* (great altar). It was dismounted from the wagon, while the priest chanted the *mantra* 'thou art the thunderbolt of Indra, killer of Vṛtra; may he (the sacrificer) kill Vṛtra through thee'. The four horses were, then, bathed and yoked to the chariot chanting the *mantras*.[45]

In the *Kātyāyana Śrautasūtra*, there is a detailed description of four horses being hatched to a chariot at the Vājapeya sacrifice. Adhvaryu, first yoked one horse on the southern side and another horse to the northern side. A third horse use then yoked to the south of the southern horse at the backside. The fourth horse reportedly walked behind the chariot unyoked. There is some confusion here, because the fourth horse should have been appropriately yoked behind the northern horse to maintain proper balance.[46] The sacrificer, then, approached the chariot, touched its sides and ascended it. The Adhvaryu followed him to the chariot, made him recite seventeen *ujjite* (victory) formulas and gave him a whip to manage the horses.[47]

In this context it may be mentioned that, the term *ratha-vāhana* is widely used in later Saṁhitās and Brāhmaṇas. It perhaps denoted a movable stand on which the chariot was placed when not in use. Weber thinks that it was used to convey the war chariot to the scene of action.[48]

The chariot appears intimately associated with wars in the Brāhmaṇas and Sūtras. The following references from the *Baudhāyana Śrautasūtra* confirm this:

> Adhvaryu yokes the four horses to the chariot infront of the sacrificial post. . . . The good charioteer standing in the chariot drives the horses wherever he desires. . . . (O priests) do you sing the praise of reins, which control (the horses) according to his mind. The horses having strong hoofs and driving the chariot, treading the foes with their toes, kill the enemies unhesitatingly. . . . (The whip) smites on their back again and again; it strikes at their thighs; O whip, do thou stimulate the alert horses in battles.[49]

Also *Kātyāyana Śrautasūtra* refers to a chariot coated with copper alloy on both sides, perhaps as a protective measure.[50]

In the *Gautama Dharmasūtra*, a ruler is advised to move around in a chariot, holding a bow.[51] In a battlefield it was unethical for a warrior to target the horses and the charioteer.[52] The vehicles of war including chariots, captured in a war were considered state property.[53]

The *Āpastamba Śulbasūtra* refers to two types of chariots, viz., *vipatha*[54] and *caraṇa.*[55] According to above commentator *vipatha* type of chariot was used on rough roads while the *caraṇa* type was different from the *krīḍāratha*.[56] There is also a view that both *vipatha* and *caraṇa* were the same[57] and it is likely that these were used for military purpose.

The *Āśvalāyana Gṛhyasūtra* relates that before the beginning of the battle the *purohita* takes a position to the west of the king's chariot. He, then, chants *mantras* from the *Ṛgveda* befitting the occasion. He also recites *mantras* over the chariot horses and also equips the king with armour, bow and arrow. It was only after these rituals that the king went about deploying his army in arrays propounded by *ācāryas* like Āditya and Uśanas.[58] In the *Vādhūla Śrautasūtra* it is said that four chariots pulled by mules should be dispatched to sixteen *janapadas* of rival kings to subdue them. In this victory march one chariot seems to have covered four *janapadas*.[59]

DIVINE CHARACTER

The chariot was treated like a divinity and was offered oblations:

1. A chariot is vigour. Through vigour, he (sacrificer) wins vigour for him. He becomes vigorous.[60]
2. Having picked out the right (*dakṣiṇa*) wheel of ones own chariot one should make an offering into its hole.[61]
3. The priest prays to the chariot as follows:

> (O chariot made of) wood, be thou of strong limbs, our companion, efficacious, great hero. Thou art fastened (with the cords of cow's sinews); be strong. May the one riding you win the (persons) to be conquered. (O sacrificer) do thou offer oblation to the chariot – the vigour derived from heaven and earth, the strength collected from trees, the vigour of waters, surrounded by the cow's (sinews), the thunderbolt of Indra. O, god chariot, do thou thunder-bolt of Indra, the face of Maruts, the embryo of Mitra, the naval of Varuṇa . . . accept our oblation.[62]

Like the chariot, horses too were also adored as a divinity. At the time of Vājapeya sacrifice, the priest caused the horses to be bathed chanting the *mantra* 'There is immortality in waters, there is healing quality in waters. O horses do you be strong by virtue of good qualities of waters'.[63] He yokes the southern horse with the *mantra* 'Vāyu, Manu and the *Gandharvas* twenty-seven in all yoked the horse first. May they install speed within thee'.[64] Similarly, he yokes the northern horse. He yokes a third horse to its north. He cleanses their backs. After the yoking of this principal chariot, the yoking of other chariots followed.[65]

Like the chariot and chariot-horses chariot-wheels were also adored. Before the beginning of the chariot race during the Vājapeya sacrifice the priest addressed the two wheels: O lord of the forest, be strong of limb, our comrade, furthering, manly; thou art tied with cowhide, be thou strong; may he who mounts thee conquer what is to be conquered.[66]

However, the divinity of the chariot is best elucidated in the *Aśvalāyana Gṛhyasūtra*:

> 'When going to mount a chariot, he (king) should touch the wheels with his two hands separately with (the words)' I touch thy two fore-feet. Thy two wheels are the Bṛhat and Rathantara (*sāmans*).
>
> 'Thy axle is the Vāmdevya' with (these words he touches) the two (naves) in which the axle rests.
>
> He should mount (the chariot) with the right foot first, with (the words), 'with Vāyu's strength I mount thee, with Indra's power and sovereignty'.
>
> He should touch the reins, or if the horses have no reins, (he should touch) the horses with a staff with (the words) with Brāhman's splendour I seize you. With truth I seize you.
>
> When (the horses) put themselves in motion, he should murmur 'Go forward to thousandfold successful vigour, divine chariot, carry us forward'.[67]

It would, thus, appear that the chariot was the predominant vehicle adored like a divinity during the times when Brāhmaṇas and Sūtras were composed, say between eighth and sixth century BC. It played a significant role in all rituals, particularly those of military significance, and was offered worship by the priests.

NOTES

1. *आत्मानां रथिनं विद्धि शरीरं रथमेव तु।*
बुद्धिं तु सारथिं विद्धि मनः प्रग्रहमेव च।।
इन्द्रियाणि हयानाहुर्विषयां स्तेषु गोचरान्। *Kaṭhopaniṣad*, 1.3.3-4; also

1.3.5-6; 1.3.9; *Bṛhadāraṇyaka Upaniṣad*, 2.5.15; *Kauśītakī Upaniṣad*, 3.8; *Śvetāsvatāra Upaniṣad*, 1.4; *Maitrāyaṇī Upaniṣad*, 2.6.

2. इमा रामाः सरथाः सतूर्या। Ibid., 1.1.25.
3. *Jaiminīya Brāhmaṇa*, 1.129-30; Also *Aitareya Brāhmaṇa*, 2.37.
4. *Aitareya Āraṇyaka*, 3.8.9.
5. *Lāṭyāyana Śrautasūtra* (*LSS*), 2.7.20-22; 2.7.26-27; 2.8.6-8; 2.8.12-13; 3.10.6-13; 5.12.9-18; 8.3.12; 8.6.9-11; 9.4.7; 9.4.13-15; Ram Gopal dates the Sūtras between 800 and 500 BC., *India in Vedic Kalpasūtras*, pp. 89-90.
6. Ibid., 2.8.6-8, 12-13.
7. Ibid., 2.7.21.
8. Ibid., 2.7.27.
9. Ibid., 9.4.13.
10. एतान्येवाश्वाश्वतरेष्वश्वतरीषु च यथार्थमूहेत्। Ibid., 2.7.26.
11. उतिष्ठ राजन परिवर्मास्यश्वयुक्तो रथो विततो दैव भारणोविशां राजा ब्राह्मणा एधि गोप्तेति। *Lāṭyāyana Śrautasūtra*, 3.10.6
12. संहत्यातिष्ठेति ब्रुयात्। Ibid., 3.10.10; It is said that in the past king's priests also served as their charioteers so that they could keep a watch that the king did not commit sin. *Jaiminīya Brāhmaṇa*, 3.94.
13. दक्षिणेन रथपथं कृत्वोत्तरेणाग्नीध्रीयं पूर्वापरे चर्माणि विवघ्नीयुः। Ibid., 3.10.11
14. *Baudhāyana Śrautasūtra*, 11.7.
15. *Mānava Śrautasūtra*, 7.1.2.32-39.
16. Ibid., 7.1.3.9.
17. Ibid., 3.7.1.3; also see *Kātyāyana Śrautasūtra*, 14.3.21.
18. *Baudhāyana Śrautasūtra*., 12.13.
19. *Kaṇva Śatapatha Brāhmaṇa*, 6.2.1-11.
20. *Lāṭyāyana Śrautasūtra*, 5.12.9-18; This is also said in *Baudhāyana Śrautasūtra* in relation to Vājapeya sacrifice. औदुम्बरं रथचक्रं सप्तदशारमिति। *Baudhāyana Śrautasūtra*, 11.1.
21. Ibid.; reference to this ceremony is also met in *Baudhāyana Śrautasūtra*, 11.1,6.
22. द्रयाणि तु खलु रथचक्राणि भवन्ति। साराणि च प्राधियुक्तानि च। *Baudhāyana Śulbasūtra*, 5.2-3.
23. Ibid., 5.38.
24. *Lāṭyāyana Śrautasūtra*, 8.3.12.
25. संस्कृतो रथः। Ibid., 9.4.7.
पुराधेयदक्षिणा ददाति पुनर्निष्कृतं रथम्। *Baudhāyana Śrautasūtra*, 3.3; The practice of presenting chariot as *dakṣiṇā* continued for centuries. Satakarṇi I, the Sātavāhana ruler, presented a 'one-horse chariot'

besides many other things after Aśvamedha *as dakṣiṇā. Comp. History of India*, vol. II, p. 302.

26. Ibid., 8.11.16; Chariots with three horses yoked to each are referred to in *Baudhāyana Śrautasūtra* (25.33) as well.
27. पृथक चतुर्युजावश्वरथावपचित्योः। *Lāṭyāyana Śrautasūtra*, 9.4.13.
28. उभयतः कांस्यकवचौ। Ibid., 9.4.14.
29. शतार्हाः स्युरश्वाः शतक्रियो वा। Ibid., 9.4.15; There are many more references to chariots, given as gifts to priests. *Lāṭyāyana Śrautasūtra*, 8.2.22-25, 8.7.3, 8.11.16, 8.11.21; *Baudhāyana Śrautasūtra*, 12.2.
30. तस्य प्रष्टिवाहनोऽश्वरथो दक्षिणा ते वै पञ्चवातास्त्रयोऽश्वा द्वौ सव्यस्थसारथी। *Kaṇva Śatapatha Brāhmaṇa*, 7.1.2.9.
31. *Baudhāyana Śrautasūtra*, 12.16.
32. *Bhāradvāja Śrautasūtra*, 5.21.1; also 14.5.2.
33. तत्र ब्रह्मणेऽश्वरथं शतञ्च दद्यात्। *Lāṭyāyana Śrautasūtra*, 8.2.22.
34. तस्य चतुर्युगश्वरथो दक्षिणा तेषामेकः श्यावः स्यात्। Ibid., 8.7.3; also *Manusmṛti*, 8.209.
35. तस्य दक्षिणा दासीनिष्करथहस्तियानगवां पृष्ट्यानाञ्चाश्वानां सप्तदशैकैकस्य जातस्य। Ibid., 8.11.16.
36. येन रथेनाजिं धावेत्तमुद्गत्रे दद्यादासन्दीञ्च सास्तरणां यामभिषिक्त आरोहेत्। Ibid., 8.11.21.
37. *Aśvalāyana Gṛhyasūtra*, 1.8.1; also *Pāraskara Gṛhyasūtra*, 1.10.1-4.
38. *Āpastamba Śulbasūtra*, 2.6.7-17; *Baudhāyana Śulbasūtra*, 1.10-12; *Kātyāyana Śulbasūtra*, 2.1-4; *Mānava Śrautasūtra*, 10.1.2.1.
39. P.C. Chakravarti, *The Art of War in Ancient India*, p. 28.
40. वैश्याचशूद्राया रथकाराः। *Baudhāyana Dharmasūtra*, 17.5.
41. *Āpastamba Śrautasūtra*, 6.1.50.
42. रथकारस्याऽधाने। *Kātyāyana Śrautasūtra*, 1.1.9.
43. Ibid.
44. *Baudhāyana Śrautasūtra*, 10.45; also *Taittirīya Saṁhitā*, 4.4.3.1-2.
45. Ibid., 11.6; This wagon was perhaps the same as रथवाहनम् of the *Ṛgveda* (6.75.8); also see *Kātyāyana Śrautasūtra*, 15.6.26-28.
46. *Kātyāyana Śrautasūtra*, 14.3.6-9.
47. *Baudhāyana Śrautasūtra*, 11.7; also 12.7; In *Pāraskara Gṛhyasūtra* (14.1-15) the mounting of chariot (*rathārohaṇa*) is described in detail.
48. *Vedic Index*, vol. II, p. 205.
49. *Baudhāyana Śrautasūtra*, 15.24.
50. *Kātyāyana Śrautasūtra*, 22.10.32.

51. *चर्या च रथधनुर्भ्याम्।* *Gautama Dharmasūtra*, 10.14.
52. Ibid., 10.17.
53. *वाहनं तु राज्ञः।* Ibid., 10.20.
54. *विपथयुगेन पुरस्तात्।* *Āpastamba Śulvasūtra*, 2.6.10.
55. *अस्थ रथश्चारण उच्यते।* Ibid., 2.6.18.
56. Ibid., pp. 308-09.
57. Ibid., p. 39.
58. *Aśvalāyana Gṛhyasūtra*, 3.10.1-19.
59. Ram Gopal, op. cit., p. 515.
60. *ओजो वै राष्ट्रभृत ओजो रथ ओजसैवास्मा ओजोऽवरुन्द्धे।* *Baudhāyana Śrautasūtra*, 14.17.
61. *स स्वरथस्य दक्षिणं चक्रं प्रवृह्य नाडीमभिजुहुयादिति।* Ibid.
62. Ibid., 15.24; also see *Taittirīya Saṁhitā*, 4.6.6.14-17; *Mānava Śrautasūtra*, 7.1.2.29, 9.1.3.17-27.
63. *Baudhāyana Śrautasūtra*, 11.6.
64. Ibid.
65. Ibid.
66. *Mānava Śrautasūtra*, 7.1.2.30; the ritual of yoking horses to the chariot is given in detail in the *Śatapatha Brāhmaṇa* also.
67. *Aśvalāyana Gṛhyasūtra*, 2.6.1-5.

CHAPTER 4

The Epic Times

THE EPIC CHARIOT

The classical Indic epics, the *Rāmāyaṇa* and the *Mahābhārata*, though composed much after the Vedas,[1] portray the ancient tradition of chariotry as in the Vedic age. S.D. Singh is perhaps right in assuming that, 'the epic war chariot is essentially the same vehicle as that of the Vedic period'.[2] The following extracts clearly portray the picture of a war chariot in the epics:

> The battle car was called *saṅgrāmika*. It bore a standard, was decked with small bells and was equipped with arrows and spears. It was worked with gold and carried a sharp edged *pratoḍa*.[3]
>
> (Chariot) large and fine, adorned with bells, with a golden net, and light in weight to run with the noise of thunder; furnished with tiger-skin, provided with *varūthina* (guard) and drawn by good fair-necked steeds.[4]
>
> Cars were drawn by four horses (*caturyuja*), and equipped with arrows and spears, and a hundred bows apiece; for each car were two pole horses fastened to the axle end (*pārṣṇi*) and driven by one driver apiece (*pārṣṇisārathi*). The battle cars were like guarded cities and horses had gilded trappings (*hemabhāṇḍa*).[5]

> He (Rāma) mounted with joy the divine war-vehicle, his bow strung, that car furnished with all kinds of weapons, sounding with a hundred bells, harnessed with tough swift steeds, and well-guided by the charioteer; which had the sound of thunder, and the glory of shining moon or sun, which had a lofty flag staff; which was irresistible, well-adorned, covered with a net of gold, on fire as it were with glory.[6]

Collating the evidence, thus furnished, Hopkins pictures the epic war chariot as follows:

Beneath is the axle (*akṣa*) to the end of which the wheels are attached, and before which is the charioteer's place, while above and behind is the place of the knight. The nest or the box above is so intimately connected with the axle that the two are often broken together (*bhagnacakrasanīḍā* Bhīṣma, 71.82). The noiseless running of the axle is especially praised. The mention of this part of the wagon often implies that the car has only two wheels.[7]

The epic war chariot (*saṅgrāmika ratha*), thus appears to be a two-wheeled light vehicle[8] pulled by two horses. Perhaps not much different from the chariots represented in Bhārhut and Sāñcī sculptures.[9] It could be lifted up by a chariot warrior, when caught in a quagmire.[10] These chariots were not very strongly made and could be shattered by arrow-shots in the battlefield.[11]

UPHOLSTERING

It appears that the war chariot was made of some light wood and its components were bound together with ropes.[12] To strengthen the body it was upholstered with some kind of skin, which in a way also served as an armour. The chariots of Kṛṣṇa,[13] Droṇa,[14] Karṇa,[15] Arjuna[16] and many Kaurava warriors were provided with a covering made of tiger skin.[17] The chariots of Alambuṣa[18] and Ghaṭotakaca[19] were covered with bear-skin. Occasionally, the skin of rhino was used.[20] Five hundred chariots of Kaurava army, which attacked Bhīmasena on the sixteenth day of war were all upholstered with lion and tiger skins.[21] Perhaps, the tiger skin was more widely used for the purpose, so as to create scare among the enemy ranks. The chariots parked at the palace of Rāvaṇa were also upholstered.[22]

It seems that leading warriors adorned their chariot with golden frames.[23] Moon and star symbols were portrayed on the body of the chariot[24] and bells,[25] flags[26] and garlands[27] added to its glamour. The chariot of Kṛṣṇa, which carried him to Hastināpur was adorned with the symbols such as, crescent, moon, fish, deer, bird, various kinds of flowers and decked with gold and jewels. The body of the chariot was upholstered with tiger skin.[28] The chariot of Indrajit, son of Rāvaṇa, too was decorated with the symbols of deer, crescent and moon.[29]

Chariot sheds (*rathaśālās*) were built within the city walls for lodging chariots.[30] Rituparṇa, a ruler of Ayodhyā, is reported to have built a *rathaśālā* in his capital.[31] Kauravas also maintained many *rathaśālās* where the chariots were safely kept. A fire in the chariot hanger was considered a bad omen.[32] In the *Rāmāyaṇa*, it is said that within the palace complex of Rāvaṇa in Laṅkā fast running chariot horses, well-decorated chariots covered with either lion or tiger-skin and the chariot warriors were lodged.[33] This is indeed a reference to *rathaśālās*.

Chariots were made in various shapes and sizes, perhaps to serve different requirements. In the epic itself, there is a specific reference to a war-chariot (*saṅgrāmika ratha*) and a luxury chariot (*krīḍāratha*).[34] Perhaps, some other types of chariots such as *deva ratha* and *rāja ratha* were also known. It seems that the epic war chariot, made of some light wood, was not a heavy vehicle and could be handled manually also. Karṇa, the Kaurava commander, tried hard to lift up his chariot when it got stuck in mud during a battle.[35] Similar to the aforesaid description, an Egyptian bas-relief of *c.*1170 BC, shows a chariot being carried on the shoulders of a man – no feat if it weighed less than hundred pounds.[36]

COMPONENTS

Various components of the chariot are specified in connection with the composition of a divine chariot by Viśvakarmā. These include *akṣa, jaṅghā, parivāra* (cover), *īṣā, yuga, kūbara, apaskara, pariskara, upaskara, anukarṣa, varūtha, triveṇu, bandhura, cakra, pūrvapakṣa, aparapakṣa, īśādaṇḍa, yoktra, yugacarma, bālabandhana, raśmi, carma, ghaṇṭā, dhvajayaṣṭi, ketu, patākā, pratoḍa, aśva,* and so forth.[37]

In the Śānti Parva, in an allegorical description the following complements of chariot are mentioned: *upastha, varūtha, kūbara, akṣa, yuga, āyu, vandhana, bandhura, nemi, vāhana, nābhi, pratoḍa, sārathi, rathī*, etc.[38]

During a fierce encounter with Saṁśaptakas, Arjuna destroyed many of their chariots.[39] The components of chariots, could be seen scattered in the battlefield which included, *kūbara, īṣā, rathacakra, yoktra, yuga, anukarṣa, koṣa, triveṇu, akṣa, jaṅghā, varūtha, adhisthāna, rathanīḍa, upastha, rathopastha, bandhura, talpa, ātapatra, chatra, dhvaja, patākā, pratoḍa,* etc.[40] During an encounter with the *gandharvas*, Karṇa suffered great humiliation as many components of his chariot lay shattered.[41] The chariot of Duryodhana too suffered a similar fate.[42] Important components of chariot included the following:

1. *Kūbara* and *īṣā*, denoted shaft or pole of the chariot.[43] It was fastened to the box of the *kāṣṭha* and to the double yoke that crossed it and rested on the neck of the steeds.[44] Hopkins explains that the lower end of the pole was called *īṣā*, while the upper-end was called *kūbara.*[45] In an encounter, Bhīṣma, when injured, is said to have held on to the *kūbara* of his chariot.[46] But the occurrence of *kūbara* and *īṣā* together[47] suggests that these were two different components.
2. *Yuga* denoted the yoke which rested on the neck of the horses. With *yoktra* (thongs) it was tied to the pole (*kūbara*) of the chariot.
3. *Rathacakra* was indeed the wheel of the chariot. *Nemi* was the rim, the outer part of the chariot-wheel. It made noise while on the move.[48]
4. *Anukarṣa* was the ballast-guard attached to the backside of the chariot to prevent it from overturning.[49] Frequent mention of *anukarṣa* in the epic suggests that it was a vital part of chariot ensemble.[50]
5. *Koṣa* denoted the box of the chariot, which provided space to sit or stand both for the warrior and the charioteer. It was often adorned with gold work.[51] *Adhiṣṭhāna* was the standing place over the wheels, while the term *pariṣkara* denoted the wheel-guard.[52]
6. *Triveṇū* has been defined as a 'threefold-piece', or 'the piece with three sticks'. It is said to have been made of bamboo wood in a triangular shape, of which one arm ran parallel to the axle and the other two arms ran together to the pole.[53] Perhaps, it was added to the chariot ensemble to lend strength and durability to the central pole.[54] Interestingly, there is also a reference to a *triveṇū* made of silver.[55]

7. *Akṣa* (axle) was the most vital component of the chariot ensemble, over which the box was raised and to which the pole was fixed. *Jaṅghā* or *akṣajaṅghā*[56] have been explained as the axle tree by Hopkins.[57]
8. *Varūtha* was a kind of protective fence around the box of the chariot. It was perhaps made of leather.[58] Occasionally, it was made of iron and adorned with gold, diamond, jems, etc.[59] *Vandhura* was the seat of the charioteer whereas *talpa* was the seat of the warrior.[60]
9. *Rathanīḍa* denoted the space on the front side of the chariot, where the charioteer was seated. In an encounter, Droṇācārya evicted the charioteer of Dhṛṣṭadyumna from *rathanīḍa*,[61] the hinder part of the car.[62] The term was occasionally used for the seat of the *rathin*.[63]
10. *Ātapatra* was an important adjunct of the chariot.[64] It protected the warrior from the sun rays and also added to the decor of the chariot. These could be seen littered on the battlefield during the epic war.[65]
11. *Dhvaja* was another important constituent of the chariot ensemble. It was raised on the hind side and comprised a *yaṣṭi* (staff), a *patākā* (flag) and a *ketu* (symbol). Hopkins has described it in some detail:

    > On the top of the staff was placed the *dhvaja* or *ketu*, the former meaning sometimes the whole arrangement, staff, image or banner, the latter the symbol or banner alone. This image was a likeness of some animal, as a boar or flamingo. Thus the *vānara* or ape-ensign of Arjuna was placed on top of the *dhvaja*, and his car was usually termed the car with ape standard.[66]

12. *Yaṣṭi* was a strong pole of wood which could bear the weight of the warrior if need be.[67] On top of it was placed the *ketu* or emblem.[68] Below the emblem, a flag was displayed so as to make the *dhvaja* visible from a distance. The flag was also displayed to augment the decor of the *dhvaja*.[69]

The Rāmāyaṇa and the Mahābhārata, all commanders are described as bearing some kind of *dhvaja* on their chariot. The *dhvajas* of some important warriors were as follows:[70]

Kapi-dhvaja	Arjuna
Mṛdaṅga-dhvaja	Yudhiṣṭhira
Siṁha-dhvaja	Bhīmasena
Śarabha-dhvaja	Nakula
Haṁsa-dhvaja	Sahadeva
Śāraṅga-dhvaja	Abhimanyu
Nāga-dhvaja	Duryodhana
Tāla-dhvaja	Bhīṣma
Nāga-kakṣya-dhvaja	Karṇa
Vṛṣabha-dhvaja	Kṛpācārya
Siṁhalāṅgūla-dhvaja	Aśvathāmān
Varāha-dhvaja	Jayadratha
Manuṣya-śīrṣa-dhvaja	Rāvaṇa
Mṛgarāja-dhvaja	Meghanāda

Chatra (parasol) was perhaps borne on the chariot by men of royal lineage. These, made in many hues looked very colourful.[71]

Ropes were necessarily used in making a chariot. In fact, various components of the chariot were tied together with ropes only.[72] *Iṣābandha, cakrabandha* and *rathabandha* were vital joints and the warriors generally targetted them to bring down the adversary.[73] It seems that the nails were not used for fixing chariot parts.

The chariot, thus constructed, was treated like a sanctified object by the *rathin*. Arjuna while mounting the chariot of Indra first circumambulated it in reverence.[74] Following the completion of Rājasūya sacrifice by Pāṇḍavas, when Kṛṣṇa left for Dvārakā in his chariot, he also performed the ritual of *pradakṣiṇā*.[75] On the fourteenth day of war the great chariot of Arjuna was sanctified with *mantras* for victory.[76] Further, he circumambulated the chariot before mounting it.[77] The chariot of Karṇa, before being driven to war was, adored by the priest to bring auspiciousness. It was then duly worshipped and circumambulated by Karṇa.[78] Similarly, Vajradaṁṣṭra, a commander of Rāvaṇa, mounted his chariot after circumambulating it.[79]

CHARIOT HORSES

The horse was basic to the concept of the chariot. In fact, it was the yoking of horse to the cart, which led to the birth of the chariot. In both the epics, chariots are generally described as being pulled either by two or four horses. The chariot of Kṛṣṇa at one place is described as being pulled by two horses (Śaibya and Sugrīva),[80] but at another place by four horses (Śaibya, Sugrīva, Meghapuṣpa and Balāhaka).[81] The chariots of Yudhiṣṭhira,[82] Dhṛṣṭadyumna,[83] Kṛpācārya,[84] Virāṭa[85] and Duryodhana[86] were each pulled by four horses. Many rulers presented chariots and horses to Yudhiṣṭhira at the time of Rājasūya sacrifice.[87] The chariots arrayed by Kaurava army in the battle of Kurukṣetra were drawn by four horses each.[88]

In the Rāmāyaṇa, the chariot of Meghanāda is described as being pulled by four horses.[89] The chariots of Khara[90] and Dūṣaṇa[91] were also pulled by four horses each. But the chariots of Rāvaṇa,[92] Durdhara[93] and Akṣayakumāra[94] were pulled by eight horses each. The chariot of Atikāya was reportedly drawn by one thousand horses.[95]

The manner in which the four horses were yoked to the chariot has been described in the epic.[96] The relative position of the horses has been explained by Hopkins as follows:

> One bears the right-hand *dhura*, one the left, the 'near' horse; one is attached to the end of the fore-axle (*pārṣṇi*) on the left; another parallel to this, to the axle end on the right. Such seems to be the arrangement according to the text, though it would not be impossible to interpret as a double span, the foremost drawing on the yoke and pole, the hinder pair on the axle.[97]

Singh, who has also dealt with the subject in some detail, however, disagrees with Hopkins stating that horses in a double span would have reduced the manoeuvrability of the chariot on a crowded battlefield.

> The universal practice attested in Indian sculpture down to the beginning of the Christian Era is that of two or four horses yoked abreast to two-wheeled vehicles. Two horses draw on the yoke on either side of the central pole; and if there are four, the out spanners are held by loose

straps attached to the ends of the axle on each side, and draw perhaps also on their neighbours' collars.[98]

Occasionally, mules[99] and asses[100] were also yoked to the chariots. The chariot of king Dhṛtarāṣṭra was pulled by mules, which could cover 14 *yojanas* in a day.[101] Rāvaṇa abducted Sītā in a chariot pulled by asses.[102] A chariot pulled by white she-mules was particularly considered a prized possession.[103] Elephant drawn chariots were also known.[104] King Drupada presented twenty-six elephant drawn chariots to Pāṇḍavas.[105] Chariots pulled by camels were also used.[106]

In the Mahābhārata the colour of the horses yoked to the chariots of important warriors has been invariably mentioned. In fact, a section in the Droṇa Parva is devoted to describe the colour of the horses yoked to their chariots[107]:

Yudhiṣṭhira	सुवर्णवर्णा	*MB*, Droṇa, 22.61
	हेमोत्तमप्रतिच्छनैर्हयै	Ibid., 22.5
Arjuna	श्वेतैर्हयैर्युक्ते	*MB*, Bhīṣma, 23.14
Bhīṣma	श्वेतैरश्वैः	Ibid., 20.9
Karṇa	सिताश्वयुक्तम्	Ibid., Karṇa, 26.58, 74
Dhṛṣṭadyumna	परावतसवर्णाश्वा	Ibid., Droṇa, 22.33, 50.89; also, Bhīṣma, 19.25
Droṇa	शोणाश्वान्	Ibid., 72.26; 7.10
Yudhāmanyu	सारङ्गाश्वो	Ibid., 23.3 (BE)
Kṣatradharmā	शोणाश्वः	Ibid., 23.5 (BE)
Drupada	ललामैर्हरिभियुक्तैः	Ibid., 22.7
Śikhaṇḍin	आमपात्रनिभाकाराः	Ibid., 22.13
Dhṛṣṭaketu	शबलैरश्वैः	Ibid., 22.16
Vṛhatkṣatra	पलालधूमवर्णाभाः	Ibid., 22.17
Kṣatradeva	पद्मवर्णा	Ibid., 22.18
Bhīmasena	ऋश्यवर्णैर्हयैः	Ibid., 22.2, 107.26
Prince of Kāśī	क्रौञ्चवर्णा हयोत्तमाः	Ibid., 22.19
Prativindhya	श्वेताः कृष्णग्रीवा	Ibid., 22.20
	मयूरसदृशैर्हयैः	Ibid., 140.11
Sutasoma	माषपुष्पसवर्णाः	Ibid., 22.21

Śatānīka	शालपुष्पनिभा हया:	Ibid., 22.23
Narasiṁha (son of Śiśupāla)	सारङ्गशबला हया:	Ibid., 22.15
Śrutakarmā	मयूरग्रीवसंनिभा:	Ibid., 22.23
Śrutakīrti	चाषपत्रनिभा हया:	Ibid., 22.25
Abhimanyu	पिशङ्गास्तं	Ibid., 23.33; Udyoga, 55.16; Bhīṣma, 100.1
Son of Vārdhakṣema	पलालकाण्डवर्णास्तु	Ibid., 22.28
Śitipāda	रुक्मपत्रैरुरश्छदै:	Ibid., 22.29
Śreṇimanta	कौशेयसदृशा हया:	Ibid., 22.30
Kāśirāja	हेमवर्णा:	Ibid., 22.31
Kuntībhoja	इन्द्रायुधसवर्णैस्तु	Ibid., 22.39
Cekitāna	वभ्रकौशेयवर्णास्तु	Ibid., 22.38
Rocamāna	अंतरिक्षसवर्णास्तु तारकाचित्रिता इव	Ibid., 22.40
Sahadeva (son of Jarāsandha)	कर्बुरा: शितिपादास्तु	Ibid., 22.41
Sudāmā	पुष्करनालस्य समवर्णा	Ibid., 22.41
Siṁhasena	शशलोहितवर्णास्तु	Ibid., 22.43
Śreṇimāna	कौशेयसदृशाहया:	Ibid., 23.37
Janmejaya	सर्षपपुष्पाणाम् तुल्यवर्णा	Ibid., 22.44
Pañcāla prince	माषवर्णाश्च	Ibid., 22.45
Daṇḍadhāra	शरकाण्डनिभा हया:	Ibid., 22.46
Vyāghradatta	रासभारुणवर्णाभा:	Ibid., 23.54 (BE)
Sudhanvā	हरय: कालकाश्चित्रा:	Ibid., 23.55 (BE)
Citrāyudha	इन्द्रगोपकसंनिभा:	Ibid., 23.56; (BE)
	किंशुकपुष्पाणां समवर्णा	Ibid., 23.64 (BE)
Sātyaki	रजताश्व	Ibid., 23.2 (BE)
	कुन्देन्दुरजतप्रभा:	Ibid., 115.21
Yudhāmanyu	सारङ्गाश्व	Ibid., 23.3; (BE) Also 191.32 (BE)
Kṣatradeva	पद्मपत्रनिभांश्चाश्वान्	Ibid., 23.6 (BE)
Nakula	शुकपत्रपरिच्छदा:	Ibid., 22.3 Udyoga, 56.15
Śrutakarmā	मयूरसदृशैर्हयै:	Ibid., 24.27
Uttamaujā	कृष्णास्तु मेघसंकाशा:	Ibid., 23.8 (BE)

Sahadeva	कृष्णास्तु मेघसंकाशाः	Ibid., 22.4 Udyoga, 55.14
Drupada	लामैर्हरिभिर्युक्तैः	Ibid., 22.7
Virāṭa	पाटलपुष्पाणां समवर्णा	Ibid., 22.9
Virāṭa's son	हरिद्रसमवर्णास्तु	Ibid., 22.10
Kaikeya princes	इन्द्रगोपकवर्णैस्तु	Ibid., 22.11
Sukṣatra	चक्रवाकोदरा हयाः	Ibid., 22.47
Satyadhṛti	शवलास्तु अश्वाः	Ibid., 22.48
Śukla	अश्वैश्च शुक्लैः	Ibid., 22.49
Satyadhṛti	अरुणाः	Ibid., 22.32
Candradeva	अश्वाः शशाङ्कसदृशाः	Ibid., 22.50
Śaibya	नीलोत्पलसवर्णास्तु	Ibid., 22.51
Rathasena	कलायपुष्पवर्णास्तु	Ibid., 22.52
Nīla	रथवाहैश्च नीलैः	Ibid., 22.55
Hemavarṇa	पुष्करपत्रस्य तुल्यवर्णा	Ibid., 22.57
Śisupāla's son	सारङ्गशबला हयाः	Ibid., 23.21 (BE)
Daṇḍaketu	श्वेताण्डाः कुक्कुटाण्डाभा	Ibid., 22.58
Pāṇḍya king	चन्द्ररश्मिनिभैर्हयैः	Ibid., 23.72 (BE)
Vṛhanta	सौवर्णस्यन्दने	Ibid., 23.77 (BE)
Jayadratha	पारावतान् सिन्धुराजस्य	Ibid., Bhīṣma, 85.34 (BE)
Meghanāda	कृष्णान् कनकभूषणान् कालाश्वयुक्ते	*Rām*, Yuddha, 89.39 Ibid., 88.2

The chariot of Kṛṣṇa was pulled by four horses. Of these Śaibya was green, Sugrīva red, Meghapuṣpa dark like cloud and Balāhaka white.[108]

The effort devoted to highlight the colour of chariot-horses may not have been meaningless. Perhaps, each of these colours denoted some kind of symbolism and may not have been simply an expression of poetic fancy as believed by Singh.[109] Interestingly, in the *Aśvaśāstram*, though a comparatively later work, a full chapter has been devoted to describe the colour of horses.[110]

HORSE ARMOUR

The horses yoked to the chariots were protected with some kind of armour. Pradyumna joined the battle with Śālva, mounting a chariot drawn by armoured horses.[111] The horses yoked to Droṇa's chariot were covered with a protective net made of gold.[112] The chariot horses of Bhīma,[113] Sahadeva[114] and Satyadhṛti[115] were similarly protected with golden nets. It is said that, chariots of Sātyaki[116] and Karṇa[117] were pulled by superior horses, protected with golden armour.

The chariot horses were more often protected by skin armour. Abhimanyu felled many chariot horses in the battlefield, despite their skin-armour.[118] Some Kaurava chariot horses were protected with the tiger skin.[119] A Pāṇḍya ruler had covered his horses with an armour made of *vaidūrya* net.[120]

It seems that the armour was washed with gold or worked with gold to look bright and beautiful.[121] If made of tiger skin, it was adorned with a moon symbol made with gold.[122] Occasionally, these were worked with gems and jewels.[123] However, the armour made of metal or skin did not provide full protection to the horse. There are numerous references to the horses felled by arrow-shots in the battlefield.[124] The harness of a chariot-horse included straps, girth, reins, birdle and the bit. These are again said to have been made of gold.[125] But the skin might have been used more appropriately for this purpose.

IMPORT OF HORSES

Indian heartland has never been known for breeding quality horses. According to Staal, horses have always been rare in India as they do not thrive on Indian soil and the climate does not smile on them.[126] The large number of horses yoked to the chariots of epic warriors, therefore, came either from peripheral areas like Sindh or from adjoining countries. In the Mahābhārata, there are repeated references to horses called Vanāyuja, Kāmboja, Parvatīya, Bāhlīka, Kaulūtaka, Ājaneya, Nadija, etc.[127] It is said that among these horses, Ājaneya came from Arabia, Bāhlīka from Bactria, and Kāmboja from Kabul. Saindhava, Kāmboja and Bāhlīka horses were preferred by the warriors.[128] The horses of Gandharvadeśa were considered to be the best as these were supposedly blessed with supernatural qualities. Citraratha, the Gandharva king, presented one hundred such horses to each of the Pāṇḍava brothers.[129] The horses of Sindhudeśa, which ran faster than the wind, were also highly prized.[130]

In this connection, it is notable that *Aśvaśāstra* of Nakula mentions more than 40 types of horses obtainable in India.[131] Among these, Kāmboja, Bāhlīka, Vanāyuja, Gandhāra, Aratta, Taittika, Yavana, Tuṣāra, Kādara and Dārada types may have been imported.

That a large number of horses were imported is also evident from some ancillary references in the Mahābhārata. During the Rājasūya sacrifice, many rulers of foreign origin presented Yudhiṣṭhira with horses and mules, perhaps considered as a prized commodity. In this connection, the following references are particularly notable:

1. Yavanas (people of Greek origin), besides other things, brought a large number of horses from the hills and these could run faster than thought.[132]
2. The people inhabiting the lands beyond Sindhu (Indus) River, including Vairam, Pārada and Kiṭaba brought many camels and asses as present.[133]
3. Bhāgdatta, ruler of Prāgjyotiṣa and the leader of the *mlechhas*, who was accompanied by Yavanas came with fast moving Ājaneya (Arab) horses.[134]
4. The rulers from the countries of Dvayākṣa, Tryākṣa, Lalāṭākṣa, Auṣṇīka, Romaka, Ekapāda, Hārahūṇa, etc., brought ten thousand horses of higher pedigree.[135]
5. Many people including Cīnas, Śakas, Oḍras, Barbaras, Vārṣṇeyas, Hārahūṇas, etc. came with a present of ten thousand (*arbuda*) mules'.[136]
6. People like Śakas, Tukhāras, Kaṅkas, Ramasa (perhaps Romaka) and Śṛṅgi brought fast moving horses.[137]

These references indicate that the demand of horses among the Indian rulers was large, that most of these demands were met by importing horses and that it were mostly the foreigners, some of whom had settled on the borderlands of India, who facilitated these imports. In this context the observations made by Pierre Ponafidine are worth quoting:

> The demand for horses turned out to be the driving force in the world economy from the early years of first millennium CE, right up to the commencement of First World War. . . . The economic and political dynamics of this exchange was dictated on the one hand by the abundant availability of good horses in the drier climates of central Asia, the Russian steppes, Iran, Arabian Peninsula and the borderlands of Turkey with a matching scarcity at least of the better breeds, in the tropical south, south-east Asia and China, on the other.[138]

The chariots pulled by asses were considered fast moving.[139] But those pulled by she-mules raced faster and could cover a distance of fourteen *yojanas* in a day.[140] If a *yojana* is considered equivalent to four or five miles a chariot could have covered a distance of about sixty miles in a day. In Greece, during the times of Homer, a chariot could cover about fifty miles between sunrise and sunset.[141]

In any case, only fast moving horses were yoked to the chariots. In the stables of Rāvaṇa many such horses were lodged.[142] The chariot of Prahasta, the commander of Rāvaṇa, was pulled by fast moving horses.[143] The horses were well-trained in various tactics related to the pulling of chariot. Even when the charioteer was killed, the horses continued to run in the desired circles, surprising every one.[144]

CHARIOT-WHEEL (*RATHA-CAKRA*)

Next to the horse, the wheel was the most important complement of a chariot. A chariot was generally provided with two wheels.[145] The chariot of Kṛṣṇa, which carried him to Hastināpur was provided with two shining wheels.[146] Vaidya, however, holds that the epic war chariot was 'always a four-wheeled construction'.[147] Rebutting this, Singh says that, 'any four wheelers in war, if they existed for the purpose, must have doubtless been few and far between, as tributes to the personal caprice of their owners'.[148] Some warriors preferred eight-wheeled vehicles. For instance, Alambuṣa[149] and Ghaṭotakaca[150] joined the war mounting chariots with eight wheels. Remarkably, there is also a reference to a chariot without wheels.[151]

It seems that, wheels were made in various shapes and sizes.[152] The wheel consisted of tire (*rathanemi*) spokes (*ara*) and hub (*nābhi*). The tire, also called fore-circle (*pramaṇḍala*), appears to have been made of iron.[153] The reference to many warriors being killed in the battlefield by *rathanemi* suggests that, some kind of scythe was attached to the wheel.[154] The warriors who protected the chariot-wheel were called *cakrarakṣaka.*[155] Their task was to ensure the protection of the commander.[156]

CHARIOT COMMANDERS (*MAHĀRATHĪS*)

The armies were led in the battlefield by the commanders mounted on chariots. The Kaurava and Pāṇḍava princes were imparted intensive training in handling chariots, and they demonstrated their skills during a show organized by Droṇācārya.[157] Yudhiṣṭhira, particularly, excelled in chariotry, but Arjuna was rated as the greatest chariot commander of all.[158] Bhīma also received intensive training in chariot-war under the tutelage of Balarāma.[159] These commanders were ranked on the basis of their reputation and performance as follows: *Rathayūthapayūthapa, atirathin, mahārathin, rathodāra, rathin, ardharathin.*

On the eve of the Kurukṣetra War, Duryodhana requested Bhīṣma to enlighten him about the ranked commanders in Kaurava and Pāṇḍava armies. Some of the listed commanders included the following:[160]

Rathayūthapayūthapa	Droṇācārya, Somadatta, Kṛpācārya, Bhūriśravā, (Kaurava)
	Abhimanyu, Ghaṭotakaca, Sātyaki, (Pāṇḍava)
Atirathī	Kṛtavarmā, Śalya, Bāhlīka, Alambuṣa (Kaurava)
	Arjuna, Purujita, Dhṛṣṭadyumna, Śreṇimāna (Pāṇḍava)
Mahārathī	Aśvatthāman, Jayadratha (Kaurava)
	Bhīmasena, Virāṭa, Drupada, Śikhaṇḍin, Dhṛṣṭaketu, Kāśirāja, Amitaujā, Cekitāna (Pāṇḍava)

Rathodāra	Duryodhana, Duḥśāsana (Kaurava) Yudhiṣṭhira, Citrāyudha, Uttara, Uttamaujā, Yudhāmanyu, Vyāghradatta (Pāṇḍava)
Rathin	Nīla, Vinda, Śakuni, Bhāgdatta (Kaurava) Nakula, Sahadeva, Kṣatradeva, Purujit (Pāṇḍava)
Ardharathin	Karṇa (Kaurava) Kṣatradharmā (Pāṇḍava)

It appears that, the number of highest ranking commanders were four on the Kaurava side, and three on the Pāṇḍava side. The number of *atirathins* on the Kaurava side were three compared to eight on the Pāṇḍava side. Among the *mahārathins*, the Pāṇḍava side had seventeen as against four on the Kaurava side. *Rathodāras* on the Pāṇḍava side numbered ten against one hundred and seventeen on the Kaurava side. All Kaurava brothers were placed in this category. The *rathins* on the Kaurava side numbered fifteen and on the Pāṇḍava side eight. There was an *ardharathin* each on both sides. *Ardharathin* was the lowest ranked commander. In his assessment, when Bhīṣma rated Karṇa as an *ardharathin*, the latter felt deeply hurt, so much so that he refused to join the battle against Pāṇḍavas as long as Bhīṣma commanded the Kaurava army.[161]

Among the Yādavas, Cārudeṣṇa, Cakradeva, Sātyaki, Kṛṣṇa, Balarāma, Sāmba and Pradyumna were *atirathīs* while Kṛtavarmā, Anādhṛṣṭi, Samīka, Samitiñjaya, Kahva, Śaṅku and Nidānta were *mahārathīs*. Two sons of Andhakabhoja and Ugrasena were also *mahārathīs*.[162] Other than the commanders, there were warriors who fought mounted on the chariots and formed part of the chariot corps. Each of these warriors was entitled to a salary of one thousand gold coins per mensum, whether engaged in war or not.[163]

THE COSTUME

The costume of a chariot warrior included *uṣṇīṣa, śirastrāṇa, varma, kavaca, antarīya, uttarīya, vastra, talatrāṇa, aṅgulitrāṇa, padatrāṇa*, etc. Warriors joined battle after putting on *saṅgrāma-sajjā* (battle dress). Bhīṣma joined battle with Paraśurāma after wearing white garments, white turban and white ornaments[164] and also a white *kavaca.*[165] Paraśurāma on the other hand, dressed in an armour, marked with symbols of sun and moon and put on a finger protector.[166] Arjuna prepared for battle after wearing *kavaca*, palm-protector and finger-protector.[167]

The armour was generally made of metal-copper, silver or iron-and worked with gold.[168] Occasionally, bronze was also used for the purpose.[169] Armour was also known as *tanuchada* and *tanutrāṇa* (body-cover and body-protector).[170] Warriors put on a helmet and turban to protect the head.[171] Karṇa joined the battle after wearing a bright helmet.[172] Duḥśāsana also donned a similar helmet.[173] The usual dress of the commanders incuded *antarīya, uttarīya, uṣṇīṣa* and some ornaments.[174] Some kind of foot-wear may also have been in use.

Senior army commanders carried a conch on their person, to render some kind of signal service in the battlefield.[175] Kṛṣṇa blew his conch (Pāñcajanya) in *ārṣava* sound asking Dāruka to come over to assist Sātyaki in the battlefield.[176] The commencement and termination of hostilities on a particular day of war was also announced by blowing conches by the commanders.[177] Among the five Pāṇḍava brothers, Yudhiṣṭhira carried the Anantavijaya, Bhīmasena the Pauṇḍra, Arjuna the Devadatta, Nakula the Sughoṣa and Sahadeva the Maṇipuṣpaka conches on their person.[178]

CHARIOTS OF GREAT EPIC WARRIORS

All wars described in the Mahābhārata and Rāmāyaṇa are completely dominated by the chariot arm. Naturally, the chariots of leading warriors have also been described in graphic details. However, most of these details are hypothetical. Chariots of some leading warriors are described below:

ARJUNA

The chariot of Arjuna was a gift to him from god Agni, who had obtained it from god Varuṇa. The horses yoked to it were born in Gandharvadeśa, were silver-white in colour and were decked with golden garland. It carried a fearsome *kapi-dhvaja* on top.[179]

This celestial chariot, created by Viśvakarmā after a long

penance, emitted shining rays and made a thunderous noise while on the move. Decorated with various kinds of jewels, it looked beautiful. Also, it bore many kinds of flags. It is said that in ancient times, the Soma defeated Dānavas mounting this chariot.[180] In the epic war, Kṛṣṇa pilotted it.[181]

Kṛṣṇa

The chariot of Kṛṣṇa was a large vehicle, pulled by four horses, viz., Śaibya, Sugrīva, Meghapuṣpa and Balāhaka. Sugrīva held up the left yoke, while Śaibya the right yoke. Behind them were yoked Meghapuṣpa and Balāhaka as *pārṣṇibāho*. These horses were of high Kāmboja pedigree.[182] The colour of Śaibya was green like parrot wing, Sugrīva was red like *palāśa* flower, Meghapuṣpa dark like a cloud and Balāhaka white.[183]

When Kṛṣṇa left for the Kaurava court on a peace mission, his chariot was equipped with necessary weapons. Its wheels looked bright like sun and moon. The body of the chariot was covered with tiger skin and decorated with the pictures of moon, fish, animals, birds, flowers and precious stones. It was also inlaid with jems and gold. The horses of the chariot were duly given a bath and appropriately decorated before being yoked.[184] The chariot-wheels made a loud noise when on move. On top of the chariot was displayed the *garuḍa-dhvaja* symbolizing the glory of Kṛṣṇa.[185] It was driven by Dāruka.[186]

Yudhiṣthira

Yudhiṣṭhira travelled in the chariot named Jaitra (triumphant). It was pulled by eight beautiful horses, of the colour of *kurar* bird, and commanded great respect. This chariot was equal to one thousand chariots in strength. While on move, it made loud noise like the roar of the sea and the thunder of the cloud.[187]

Covered with tiger skin it was mounted on beautiful wheels. Nets of small bells decordated it on all sides.[188] It is said that, the chariot was decorated with gold and pulled by white horses with black tail, which ran at the speed of thought.[189] Yudhiṣṭhira lost this chariot to Śakuni in a gambling bout.[190]

INDRA

Driven by Mātali, the chariot of Indra was pulled by ten thousand horses of white and yellow colour. It thundered like clouds when on move. It was equipped with all kind of weapons like sword, spear, mace, etc., as also heavy stone balls. On top of the chariot, flew the bright blue coloured Vaijayanta.[191] Mounting this chariot, Indra is said to have defeated Sambāsura, Namuci, Bala, Vṛtra, Prahlāda and many more *daityas.*[192]

Indra provided Arjuna with this chariot to fight the Nivāta-kavacas.[193] During the battle of Laṅka, Indra sent charioteer Mātali along with this chariot to assist Rāma against Rāvaṇa.[194]

JARĀSANDHA

The name of Jarāsandha's chariot, which was captured by Pāṇḍavas after killing him, was Sodaryavāna.[195] It was beautiful, unconquer-able and provided space for two warriors to fight on.[196] It was bright like purified gold and decorated with small bells and frills. When on move, the chariot thundered like clouds.[197] Divine horses pulled this chariot and on top of it flew the god-created standard, which was multi-coloured and bright like Indra's bow and could be seen from a *yojana* long distance.[198]

Sodaryavāna originally belonged to Indra, and he is said to have killed ninety-nine *dānavas* mounting this chariot.[199] It was gifted to king Vasu by Indra and then passed on to kings Bṛhadratha and Jarāsandha respectively.[200] Later, it was presented to Kṛṣṇa by Yudhiṣṭhira.[201]

BHĪṢMA

The pleasing chariot of Bhīṣma was made of silver and was pulled by white horses. It was fully equipped, provided for comfortable seats and was protected with tiger skin. It was endowed with power-ful weapons and all kinds of implements. A charioteer, who was brave, well trained, a noble-man and an expert in veterinary science (*hayaśāstra*) directed and controlled this chariot.[202]

The horses yoked to this chariot moved at the speed of mind and wind and it took the charioteer little time to carry Bhīṣma to the battle arena of Kurukṣetra from Hastināpur to challenge Paraśurāma.[203] This great chariot was mounted with a *dhvaja* bearing a crest of five blue stars.[204] At another place, it is said that the *dhvaja* of Bhīṣma comprised sun and stars symbols.[205] Still in another context, it is said to have comprised a palm tree and five stars.[206]

Droṇācārya

The chariot of Droṇa was built in accordance with the prescribed rules and designed like the city of *gandharvas*. Its fluttering flag immensely delighted the heart of the warrior. Mounting this chariot, Droṇācārya destroyed the enemy armies.[207] The chariot was pulled by red horses capable of running fast like the wind. These horses were of Ājaneya pedigree and could cover long-distances leaping ahead without any rest.[208]

It was made of gold and protected with tiger skin on all sides.[209] On top of the chariot was displayed a golden *dhvaja* marked with the black dear-skin and a water-jar (*kamaṇḍalu*).[210]

Karṇa

The chariot of Karṇa was beautiful, decorated with golden garlands and inlaid with jewels shining like sun and moon. It was covered with tiger skin and well-stuffed with weapons and equipment.[211] When on move, it thundered like clouds.[212]

To this chariot were yoked four white horses,[213] which resembled bright white clouds. They were strongly built, fast running and decorated with golden ornaments.[214] The chariot was mounted with a *dhvaja* comprising a golden elephant-chain and a lotus.[215] It was also decorated with a garland of precious jewels.[216]

Bhīma

The chariot of Bhīma was pulled by horses, which looked like bear in colour.[217] These horses ran fast like mind and wind and were well-trained. The charioteer named Viśoka handled these horses.[218] On top of his chariot was displayed the *simha-dhvaja* of golden colour.[219] The eyes of the *simha* were made of *vaidūrya maṇi*.[220]

Ghaṭotkaca

The chariot of Ghaṭotkaca was a large vehicle with eight wheels. It was decorated with hundreds of small bells, which emitted melodious tunes. The chariot was covered by bear-skin on all sides and was fully stocked with weapons. While on move, it made noise like thundering clouds.[221]

This big chariot was pulled by one hundred horses, resembling the intoxicated elephants. Their eyes were red and they could run at any speed as desired.[222]

On the top of the chariot flew a red sky-high flag surmounted by a fearsome red-headed vulture.[223] The charioteer named Virūpākṣa managed the horses of the chariot.[224]

Sātyaki

The chariot of Sātyaki was inlaid with a thousand stars and the net of bells attached to it made a sweet sound. It was stocked with weapons and equipment. Enroute, it thundered like clouds.

This chariot was pulled by Saindhava horses, white like the moon. These horses of a higher breed were protected with golden armour and decorated with golden ornaments. They could run at the speed of sound and wind.[225]

On top of the chariot was displayed the *simha-dhvaja* and the flag.[226] The charioteer of Sātyaki was the brother of Dāruka, the charioteer of Kṛṣṇa, and he was an expert in preparing a chariot for war.[227]

Pradyumna

Pradyumna, son of Kṛṣṇa, joined the battle with Śālva king mounting a chariot made of gold. It was pulled by armoured horses, which moved at a very high speed as if flying in the sky. On top of the chariot, the *makara-dhvaja* was displayed.[228]

The son of Dāruka, who managed this chariot in war, was a well-trained charioteer.[229]

Rāvaṇa

Before joining the battle with Rāma, Rāvaṇa asked his men to prepare his chariot.[230] The chariot was decked with jewelled pillars, golden jars, various ornaments and nets of small bells. It was also stocked with weapons.[231] While on move, it thundered like clouds.[232]

The chariot was bright like the sun and was pulled by eight black horses.[233] The standard on top of the chariot symbolised a human-head.[234]

Indrajit

The chariot of Indrajit was resplendent, all made of gold. Its body was marked with the symbols of deer, and half and full-moon, which made it look very attractive. It was equipped with weapons like *brahmāstra* and was imbued with the power to disappear at will.[235]

The *dhvaja* raised on the chariot was bright like the fire and was decorated with golden bangles and sapphires.[236] It was pulled by four black horses decorated with ornaments of gold.[237]

When this chariot of Indrajit was destroyed by Lakṣmaṇa in an encounter, he quietly slipped out and brought a new chariot, which was again well-decorated and stocked with weapons and driven by a knowledgable charioteer.[238]

Atikāya

Atikāya, the son and a ranked commander of Rāvaṇa, was carried by a large chariot pulled by one thousand horses.[239] It was driven by four charioteers and thundered like clouds while on move.[240] Necessary weapons were also stocked in the chariot.[241] On top of the chariot was displayed the *dhvaja* marked by the symbol of Rāhu as also some flags.[242]

A VALUED GIFT

During the epic times the chariot was considered a highly valued commodity and a prized gift item. After the Pāṇḍavas had married

Draupadī, Kṛṣṇa sent them a large number of gifts including elephants and chariots.[243] Post-marriage of Subhadrā and Arjuna, when Kṛṣṇa and Vṛṣṇi notables came to Indraprastha, they brought with them a huge dowry including one thousand chariots, worked with gold and pulled by four horses each.[244] During the Rājasūya sacrifice, many rulers presented Yudhiṣṭhira with chariots of various types (*ākāra*) worked with gold, upholstered with tiger skin and pulled by well-trained horses.[245] When Kṛṣṇa came to Kaurava court as an emissary of Pāṇḍavas Dhṛtarāṣṭra proposed to present him sixteen golden chariots, each yoked to four horses of Bāhalīka pedigree.[246] Chariots were also given as a reward.[247]

It seems that elephant-driven, horse-driven and ass-driven chariots were given as gift to priests.[248] In the *Mahābhārata*, there is a reference to a priest going to attend a sacrificial ritual mounting a chariot pulled by asses.[249] Occasionally, the Brāhmaṇas preferred to travel in *saṁgrāmika-ratha* instead of *krīḍāratha*, which was perhaps more common mode of transport.[250]

CHARIOT ARM

The commanders alone did not fight mounted on a chariot. The chariot corps formed an integral part of *caturaṅgiṇī* (fourfold army) and was respected as a senior arm. Yudhiṣṭhira maintained a crack force of one thousand chariots on regular basis. Each *rathin* of this force was paid a salary of one thousand gold coins per mensum, both during war and peace.[251] If the value of one gold coin is rated at 100 rupees the pay of a charioteer may be calculated as one lakh rupees per mensum. Further, thirty thousand chariots followed him during his prosperous times.[252] The *rathins* were recruited in the chariot corps after testing their skills and the salary was also fixed accordingly.[253]

Sudakṣiṇa, the ruler of Kāmboja, joined the Kaurava camp on the eve of the war with a powerful chariot corps.[254] The ruler of Mahiṣmati also joined the Kauravas with a powerful chariot army.[255] The army of king Virāṭa joined battle against Trigartas with a large force, including ten thousand chariots.[256] The Kalinga army joined the epic war with a strong component of chariot arm.[257] Trigarta king Suśarmā joined the battle with an army of thirty thousand chariots.[258] His brothers also came with ten thousand chariots, while some friendly countries contributed ten thousand more chariots.[259] King Jayadratha deployed his army, including one lakh chariots, in the forefront of Kaurava battle array.[260]

On the second day of war, Pāṇḍavas formed *krauñcāruṇa-vyūha* and deployed a massive chariot force. They placed ten thousand chariots on the wings, one lakh on the head, ten crore and twenty thousand on the back and one lakh and seventy thousand on the neck.[261] The figures, though far from the truth, however, suggest the importance attached to the chariot arm during the epic times. It is notable that in the epic war, a total of 18 *akṣauhiṇīs* participated – eleven on the Kaurava side and seven on the Pāṇḍavas side. The chariot component of an *akṣauhiṇī*, which also included elephants, horses and foot-men, has been given in Table 4.1:[262]

	Chariots	Elephants	Horses	Foot-men
Patti	1	1	3	5
Senāmukha	3	3	9	15
Gulma	9	9	27	45
Gaṇa	27	27	81	135
Vāhinī	81	81	243	405
Pṛtanā	243	243	729	1,215
Camū	729	729	2,187	3,645
Anīkinī	2,187	2,187	6,561	10,935
Akṣauhiṇī	21,870	21,870	65,610	1,09,350

The said *akṣauhiṇī* figures when multiplied by eighteen, i.e. the number of *akṣauhiṇīs* deployed in the epic war, would suggest that a total of 393,660 chariots, 393,660 elephants, 1,080,980 horses and 1,968,300 foot-soldiers were deployed in the battlefield of Kurukṣetra. These figures are indeed imaginary and may have to be divided 'by a hundred or thousand to reach a reasonable limit'.[263] The absurdity of these figures is further augmented when compared with the remnant armies (*vale śiṣṭam*) on the last day of war.[264] On comparison more people appear in the battlefield at the end of the war than at the beginning.

The *Rāmāyaṇa* also gives highly exaggerated figures of the troops deployed by Rāvaṇa during the battle of Laṅkā. His commanders are said to have arrayed an army of more than one lakh chariots, three lakh elephants, 60 crore donkeys and camels, 60 crore horses and innumerable foot soldiers.[265] At another place the chariot corps of Rāvaṇa in stated to be 10 thousand strong.[266] The army maintained by King Daśaratha of Ayodhyā included chariots also.[267] Bharata was accompanied by a large chariot force during his visit to Rāma in the forest.[268]

In any case, chariots played a decisive role in the epic battles. In an encounter Arjuna and Bhīmasena attacked Droṇācārya on both flanks with a large chariot force.[269] Kaurava prince Durmarṣaṇa joined battle with Arjuna, with an army including 1,000 chariots.[270] The army of Laṅkā which was annihilated by Rāma included a large number of chariots.[271]

It seems that a set pattern was followed in the deployment of chariots in the battlefield. Bhīṣma advised Yudhiṣṭhira to deploy the chariots between the elephants ahead of the horses, and place the foot soldiers between the horses. He emphasized that a king who deployed his troops in the said manner always conquers his enemies.[272]

THE CHARIOTEER (*SĀRATHI*)

The victory of a warrior in the battle depended on the skill of his charioteer variously called *sūta, sārathi, yantar, niyantar*, etc. Persons, well-versed in weapon system, born in respectable families and knowledgeable on the pedigree of the horses were appointed as *sārathi*.[273] One thousand chariots, presented by Kṛṣṇa to Pāṇḍavas as dowry after the marriage of Subhadrā were driven by well-trained *sūtas*.[274] Mātali, the *sarathi* of Indra, excelled in the management of horses.[275] The *sārathi* of Bhīṣma had mastered the treatise on horse (*hayaśastra*), related to the handling of horses in battle.[276]

The charioteer of Pradyumna, who was a son of Dāruka, was again a well-trained person.[277] Arjuna himself claimed to be proficient in the art of chariotry.[278] A charioteer was expected to be proficient in wielding weapons, loading arms, providing medical care and in interpreting omens.[279] In the Kaurava army, only such persons were engaged as charioteers.[280]

The warriors were also trained to manage the chariot in the battle-field. Uttara, son of King Virāṭa, was well trained in chariotry.[281] Duḥśāsana and Abhimanyu were also well-versed in the art of chariot management.[282] Needless to say that, Kṛṣṇa and Śalya were the greatest charioteers of their times.[283] Indrajit, son of Rāvaṇa, took up reins himself after his charioteer was killed by Lakṣmaṇa in an encounter. Fighting, while keeping horses under control, was an astonishing feat indeed.[284]

Generally, one to three charioteers were engaged to drive a chariot. According to Hopkins,

> the number of charioteers depended on the horses. When two horses are sufficient, one *sārathī* or a charioteer is sufficient also. In the case of four horses (two fastened to the pole, two by straps outside, not tandem: *dhur* and *pārṣṇi*), we have one charioteer in the middle, who guides the pole horses and on each side of him the two drivers of the outer steeds.[285] There are many references to *pārṣṇi sārathi* in the epic.[286]

But Hopkins viewed that, two drivers were engaged to drive the two outer horses perhaps needs revision because the use of 'तौ' in the said *śloka* suggests the presence of only two and not three charioteers, one for the pole horses and one for the outer horses.[287] There is a reference to the killing of two *sārathīs* of Añjanparvā by Aśvatthāman with two arrow-shots. This suggests that Añjanparvā's chariot was driven by two charioteers, one *sārāthī* and one *pārsṇi sārathī*.[288]

However, chariot was generally managed by one charioteer. Even the gods engaged only one charioteer to manage their chariots. Aruṇa, son of Vanitā, was the charioteer of god Sūrya.[289] Mātali drove the chariot of Indra.[290] Dāruka was the charioteer of Kṛṣṇa.[291] The chariot of Arjuna was driven by Kṛṣṇa alone.[292] Sañjaya was the famous charioteer of Dhṛtarāṣṭra.[293] The chariots of Yudhiṣṭhira, Karṇa, Bhīma and Abhimanyu were driven by Indrasena, Śalya, Viśoka and Sumitra respectively.[294] Uttara, the son of Virāṭa, also served as the *sārathī* of Arjuna during an encounter.[295] Kīcaka, the *senāpati* of Virāṭa, also served as his charioteer.[296]

Though the men of status like Kṛṣṇa, Śalya, Uttara and Sañjaya served as charioteers generally the office of *sārathī* did not command much respect. It was for this reason that Śalya agreed to serve as the charioteer of Karṇa after considerable persuasion by Duryodhana.[297] Karṇa was looked down upon by Pāṇḍavas, because he was known to be a *sūta* by birth, and Bhīma did not consider him even fit to be killed by Arjuna for the same reason.[298]

The main function of the charioteer was to manoeuvre the vehicle tactfully in the battlefield to defeat the enemy and ensure the safety of the warrior. According to Hopkins, 'his true art consisted in wheeling and turning, in bringing the car rapidly about, so as to attack the antagonist with such speed from all quarters that the chariot seemed to advance from all sides at once'.[299] When Pradyumna fell unconscious during a battle with Śālva king, his *sārathī* removed him to a safe side.[300] During an encounter with Paraśurāma, when Bhīṣma fell unconscious, the *sārathi* drove him away from the battlefield.[301] In the battle of Laṅkā, the charioteer removed injured Rāvaṇa from the battlefield.[302] During a fierce encounter between Arjuna and Bhīṣma, Kṛṣṇa displayed exceptional skill in handling the horses to foil the arrow shots of the latter.[303] In a battle with Bhīma when Karṇa fell unconscious, Śalya removed him to safety.[304] In an encounter between Sātyaki and Karṇa, Dāruka while handling the chariot of the former, conducted many moves which astonished the onlookers.[305]

There is an interesting reference to the duties of a charioteer in the *Rāmāyaṇa*. When the *sārathi* of Rāvaṇa removed him from the battlefield in an injured state, the latter expressed his extreme displeasure. The *sārathi* pleaded that he was compelled to remove the chariot because the master had got exhausted and horses had tired. He further explained that it was necessary for a *sārathi* to understand the factors like space and time, favourable and unfavourable indications and also the strong and weak points of the warrior. He should also be aware of the terrain, suitable time for battle and weaker points of his army. Further, he should also know when to attack, when to stand firm and when to withdraw from the battlefield.[306]

The charioteer often served as an advisor to the warrior in the battlefield. Kṛṣṇa repeatedly advised Arjuna in the war of

Kurukṣetra.[307] During the battle with Śālva, Dāruka advised Kṛṣṇa to kill the former as his presence was dangerous.[308] During an encounter, when Abhimanyu asked his charioteer Sumitra to advance against the army of Droṇa, the latter asked him to think over the matter coolly as Droṇa was a great warrior and he was still a novice in the art of war.[309] Indrajit engaged a charioteer who knew all about the pedigree of horses and was capable of tendering appropriate advice.[310]

In performing his duties a charioteer even risked his life. During an encounter with Paraśurāma the *sārathi* of Bhīṣma was killed.[311] Bhīmasena, in a combat, killed the *sārathi* of Duryodhana along with four chariot-horses.[312] Another *sārathi* of Duryodhana was killed by Dhṛṣṭadyumna.[313] In the epic there ware many more references to the charioteers being killed in the battlefield.[314]

In the event, the charioteer became a casualty, the warrior himself picked up the reins. Bhīṣma himself managed his horses when his *sārathi* died.[315] When Bhīmasena killed the *sārathi* of Droṇācārya, the latter himself took up the reins in his hands.[316] Duḥśāsana also took up the reins after the death of his charioteer.[317] In the battle of Laṅkā, when Lakṣmaṇa severed the head of Indrajit's *sārathi*, the latter himself managed the chariot.[318]

The charioteer managed the horses with the help of a rein (*raśmi*)[319] and a hunter (*pratoḍa*).[320] While *raśmi* helped him to drive the horses in the dersired direction, *pratoḍa* helped him to discipline them. Falling down of *pratoḍa* from the hand of *sārathi* was considered a bad omen.[321]

The person who served as yoker or harnesser of chariot was called *rathayojaka*. Yuddhiṣṭhira before joining battle with Śalya desired that his chariot should be fully equipped with weapons and implements as laid down in the *Śāstras* by *rathayojaka*.[322] Obviously, equipping a chariot was a specialised job.

Importance of a Charioteer

The importance of the office of charioteer is best described in the *Karṇa* Parva of the *Mahābhārata*.[323] Karṇa after assuming the charge of Kaurava *senāpati* told Duryodhana that despite his being superior to Arjuna in bowmanship, the latter has the advantage of Kṛṣṇa as his *sārathi* and a divine chariot. Therefore, if Śalya, 'who was equal to Kṛṣṇa in the art of chariotry, could serve as my *sārathi* your victory is cartain'.[324]

During Tripura-*vadha* when Śiva led the army of the gods, mounting a divine chariot, Brahmā agreed to serve as his charioteer.[325] In fact, Śiva had himself asked the gods to search for *a sārathī* who was superior to him.[326] It may be recalled here that when offered the choice of selecting either Kṛṣṇa or his Nārāyaṇī *senā* of 10 crore *gopas*, Arjuna preferred Kṛṣṇa.[327] It was indeed, a long cherished desire of Arjuna to engage Kṛṣṇa as his charioteer.[328]

Famous charioteers who figure in the epics include the following:

Kṛṣṇa	The ruler of Dvārakā. He served as the charioteer of Arjuna (Udyoga, 7.34-5).
Śalya	The ruler of Madradeśa. He served as the charioteer of Karṇa during the epic war (Karṇa, 32).
Mātali	The charioteer of Indra. He drove Arjuna in his battle against Nivātkavacas and Rāma in his battle against Rāvaṇa (Āraṇyaka, 169.11, 274.15).
Dāruka	The charioteer of Kṛṣṇa. His brother and son also served as charioteers of Sātyaki and Pradyumna respectively (Āraṇyaka, 9).
Kīcaka	The Commander-in-Chief of King Virāṭa, who also served as his charioteer (Virāṭa, 25.20).
Sañjaya	The charioteer of Dhṛtarāṣṭra, the Kaurava king (Āraṇyaka, 48.1).
Nala	The Naiṣadha king who excelled in the art of driving horses (Āraṇyaka, 69).
Indrasena	The charioteer of Yudhiṣthira (Vana, 255).
Sumitra	The charioteer of Abhimaṇyu (Droṇa, 35.3).
Viśoka	The charioteer of Bhīṣma (Bhiṣma, 77.19).
Virūpākṣa	The charioteer of Ghaṭotkaca (Droṇa, 175.15).
Uttara	The son of King Virāṭa. He served as charioteer of Arjuna during a battle (Virāṭa, 45.18-19).
Aruṇa	The charioteer of God Sun (Droṇa, 175.16).
Akṛtavarṇa	The charioteer of Paraśurāma (Udyoga, 180.9).

CAKRA-RAKṢAKA (PROTECTOR OF WHEELS)

Leading commanders were protected by noted warriors on the flanks or the right and left wheels in deciding engagements. Arjuna, while shielding Śikhaṇḍin against the onslaught of Bhīṣma was protected on the lẹft side (*vāmacakram*) by Yudhāmanyu and on the right side (*dakṣiṇacakram*) by Uttamauja.[329] These commanders again protected his wheels when he launched an attack an Jayadratha.[330] On the Kaurava side, Jayadratha was protected by Aśvatthāman on the right flank and Karṇa on the left flank against the attack of Arjuna.[331] Yet, in another encounter, Duryodhana deployed Kṛtavarmā to protect the right wheel and Śalya the left wheel of Droṇācārya.[332] Later, during an encounter with Śalya Yudhiṣṭhira desired that Dhṛṣṭadyumna and Sātyaki protect his left and right wheels respectively, while Arjuna supported him from behind.[333]

TACTICS OF CHARIOT WAR

There was a method in deploying chariots in the battlefield. To start with the following measures were taken:[334]

1. Persons who were brave (*śūrāh*), knew the pedigree of the horses (*hayayonijñaḥ*) and belonged to respectable families (*kulīna*) were appointed charioteer (*sārathi*).
2. All components of the chariot were strongly tied with ropes so that they did not fall apart. Standard and flags were tied to it, besides the other adornments.
3. Two or four horses were yoked to a chariot and each of the two horses were protected by one guard.
4. Each chariot was equipped with weapons like sword, shield, bow, etc. Necessary implements and medicines were also stocked in the chariot for emergency use.
5. It seems that chariots were organized in units during a battle. In an encounter with *dānavas*, Arjuna saw their chariots arrayed in groups of hundreds and thousands.[335]

Thus, prepared and equipped the chariot corps entered the battle arena. There are numerous references to the tactics adopted by the charioteers in the battlefield. Virāṭa defeated a sizeable Trigarta army by applying different tactics of chariot war.[336] During an encounter with Kṛpācārya, Uttara, the charioteer of Arjuna, manoeuvred his chariot from the right side following the tactics of *yamakamaṇḍala* so as to evade the arrow-shots of the former.[337] In a battle with Saṁsaptakas, Kṛṣṇa manoeuvred the chariot of Arjuna forward and backward in circles to keep the adversary at bay.[338] In the fierce encounter between Droṇa and Dhṛṣṭadyumna, both warriors adopted *maṇḍala, yamaka* and some other tactics of war to defeat each other.[339]

There is a graphic description of a chariot combat between Aśvatthāman and Bhīmasena. In the fierce encounter, Aśvatthāman managed to corner Bhīma to his right and showered arrows on him.[340] As a counter measure Bhīma applying *gata* (advance) and *pratyāgata* (fall back) tactics of the *maṇḍala* (encirclement) strategy was able to corner Aśvatthāman.[341] Thus both warriors tried to unseat and kill each other by following the *maṇḍala* strategy of chariot combat.[342]

It seems that *maṇḍala* was the most widely used tactic in chariot encounters. It was applied in war in three forms, viz. *uttama, madhyama* and *adhama.* Men with intensive training in chariotry like Kṛṣṇa alone could use these tactics.[343] Bhīma used *ardha-maṇḍala* tactics against Karṇa in an encounter.[344] Dāruka, while driving the chariot of Sātyaki in a battle against Karṇa displayed many tactics including *gata, pratyāgata, āvarta, maṇḍala* and *saṅivartana.*[345] The son of Dāruka too was well-versed in various tactics of chariot war.[346]

Outflanking of the adversary was a widely used tactics. When Nakula outflanked Duryodhana on the left, the latter quickly moved to force Nakula to his right. But, Nakula foiled the attempt as he was quite familiar with the tactics of war.[347] Outflanking tactics were also applied by Abhimanyu and Duḥśāsana against each other.[348] In the battle against Śālva, the charioteer of Pradyumna also applied these tactics successfully.[349] Occasionally, the tactics of partial-encirclement (*ardha-maṇḍala*) were adopted.[350]

Koṣṭhakī (art of total encirclement) of the enemy was applied to overwhelm the adversary. When Bhīmasena was grievously injured by Kṛtavarman in an encounter, the Pāṇḍavas used this tactics against him.[351] In another encounter, Kauravas encircled Sātyaki with the intent of killing him.[352] The other starategies related to chariot war were known as *rathamāyā* and *rathacaryā*.[353] Perhaps, these involved rapid movement of the chariot to confuse the enemy.

Rathaśakti was a kind of weapon made of iron. It was perhaps used by chariot warriors only. In an engagement Duḥśāsana hurled it on Bhīma, but the latter destroyed it.[354] During the battle of Laṅkā Jambumāli attacked Hanumān on his chest with *rathaśakti*.[355]

Encounter between two chariot warriors was called *dvairatha yuddha*. The Trigarta ruler Suśarmā and Virāṭa encountered each other in a *dvairatha yuddha*.[356] There are many more references to this kind of combat in the epic.[357] The chariot arm could be best used in a battlefield which was not marked by mud and pits.[358]

Briefly speaking the following tactics related to chariot war are mentioned in the epic:[359]

1. *Maṇḍala*	Circle
2. *Ardha-maṇḍala*	Half-circle
3. *Yamaka*	Twofold, double
4. *Yamaka-maṇḍala*	double encirclement
5. *Gata*	Advance
6. *Pratyāgata*	Fall back
7. *Āvarta*	Turn around
8. *Saṁnivartana*	Turn back
9. *Sabya*	Encirclement from the left flank
10. *Apasabya*	Encirclement from the right flank
11. *Dakṣiṇa*	Encirclement of the southern flank
12. *Koṣṭhakī*	Total encirclement

It would thus appear that, during the epic times chariot dominated the Indian battlefield, perhaps much more than in the Vedic times. But by now, the other three arms, viz. infantry, cavalry and the elephant corps had also gained in importance, thus leading to the concept of *caturaṅgiṇī*. The composition of an army (*akṣauhiṇī*), then included 21,870 chariots, 21,870 elephants 65,610 horses, and 1,09,350 foot soldiers. Further, barring some exceptions, the kings and commanders fought mounted on chariots only. It is note-worthy that by now many strategies relating to the deployment and use of chariots in war had been perfected by the ancient Indians.

MEANS OF TRANSPORT

During the epic times, chariot was widely used is a mean of transport. The following illustrations prove this beyond any doubt.

When Yudhiṣṭhira desired to consult Kṛṣṇa on holding the Rājasūya ceremony, he dispatched Indrasena, his charioteer, to Dvārakā, in a fast-moving chariot to bring him to Indraprastha.[360] Kṛṣṇa rushed to the Pāṇḍava capital with Indrasena passing through many kingdoms on the way.[361]

After Subhadrā and Arjuna's marriage, Kṛṣṇa came to Pāṇḍava capital with a huge dowry, travelling in his chariot driven by Dāruka. On his return journey to Dvārakā the Pāṇḍavas, accompanied him for half a *yojana*, Yudhiṣṭhira holding the reins of his chariot and Arjuna the flywhisk,[362] as a mark of respect.

After the completion of Rājasūya *yajña*, Kṛṣṇa left for Dvārakā in his chariot. He circumambulated the chariot before mounting it. Pāṇḍavas walked some distance to see him off.[363]

After agreeing to *dyūtakrīḍā* (gambling) between Duryodhana and Yudhiṣṭhira, Dhṛtarāṣṭra ordered Vidura to go to Khāṇḍavaprastha in a fast running chariot to invite Yudhiṣṭhira for the game.[364] Pāṇḍavas travelled to Hastināpura mounting the royal chariot named Jaitra.[365] In addition, a large chariot force also accompanied them.[366]

In the *Rāmāyaṇa*, there is a detailed description of Bharata's journey from Kekaya kingdom to Ayodhyā. He moved at a very fast pace negotiating many chiefdoms, rivers and forests mounting his chariot.[367]

During the journey to Citrakūṭa, the three queens and many more people travelled by chariots. Besides that an army of sixty thousand chariots also accompanied Bharata.[368]

NOTES

1. It is generally believed that the *Mahābhārata* obtained its present shape between 400 BC and 400 AD while the final compilation of the *Rāmāyaṇa* was done between 200 BC and 200 AD.
2. S.D. Singh, *Ancient Indian Warfare*, p. 36.
3. *सज्जीकुरु रथं क्षिप्रं यस्ते संग्रामिको मतः।*
   सायुधः सपताकश्च सशक्तिः कणयष्टिमान्।।
   किङ्किणीशतनिर्घोषो युक्तस्तोमरकल्पनैः।
   गदाखड्ग निबद्धश्च परमेषुशतान्वितः।। *MB* (*Mahābhārata*), Anuśāsana, 53.29-30;
   त्रिदंष्ट्रं वज्रसूच्यग्रं प्रतोदं तत्र चादधत्। Ibid., 53.32.
4. *ततो रथेन शुभ्रेण महता किङ्किणीकिना।*
   हेमजालविचित्रेण लघुना मेधनादिना।।
   सुपस्करेण शुभ्रेण वैयाघ्रेण वरूथिना।
   सैन्यसुग्रीवयुक्तेन प्रत्यदृश्यत दारुकः।। *MB*, Udyoga, 129.21-2.
5. *कुलीना हययोनिज्ञाः सारथ्ये विनिवेशिताः।*
   बद्धारिष्टा बद्धकक्ष्या बद्धध्वजपताकिनः।।
   चतुर्युजो रथाः सर्वे सर्वे शस्त्रसमायुताः।
   संहृष्टवाहनाः सर्वे सर्वे शतशरासनाः।।
   धुर्ययोर्हययोरेकस्तथान्यौ पार्ष्णिसारथी।
   तौ चापि रथिनां श्रेष्ठौ रथी च हयवित्तथा।। Ibid., 152.9-11.
6. *ततः काञ्चनचित्राङ्गः किङ्किणीशतभूषितः।*
   तरुणादित्यसंकाशो वैदुर्यमयकूबरः।।
   सदश्वैः काञ्चनापीडैर्युक्तः श्वेतप्रकीर्णकैः।
   हरिभिः सूर्यसंकाशैर्हेमजालविभूषितैः।।
   रुक्मवेणुध्वजः श्रीमान् देवराजरथो वरः।
   देवराजेन संदिष्टो रथमारुह्यमातलिः।।
   अभ्यवर्तत काकुत्स्थमवर्तीय त्रिविष्टपात्।
   Rām (*Rāmāyaṇa*) Yuddha, 102.10-12.
7. E.W. Hopkins, *Social and Military Position of the Ruling Caste in Ancient India*, p. 181.
8. *चन्द्रसूर्यप्रकशाभ्यां चक्राभ्यां समलंकृतम्।* *MB*, Udyoga, 81.15; Karṇa, 24.71; Anuśāsana, 53.27-9.
9. A. Cunningham, *The Bhārhut*, Pls XIII.3, XXV.4; J. Marshall, and A. Foucher, *Sāñcī*, Pls. 11.2, 15.3, 18a.2, 35a.1, 35b, 50a.1, 61.2, 62.2, etc.
10. *MB*, Karṇa, 66.64.
11. There are many references to shattered war chariots in the epic.
12. *ईषावन्धं चक्रबन्धं रथबन्धं तथैव च।* *MB*, Droṇa, 164.132.
    बद्धारिष्टा बद्धकक्ष्या बद्धध्वजपताकिनः। Ibid., Udyoga, 152.9

13. *वैयाघ्रेण वरूथिना।* Ibid., Udyoga, 129.22.
14. *वैयाघ्रपरिवारणान्।* Ibid., Droṇa, 8.8; also Sabhā, 47.29.
15. Ibid., Karṇa, 56.11; also 37.27; Udyoga, 81.18.
16. *रथं वैयाघ्रं।* Ibid., 50.39.
17. *सुवर्णचित्रैर्वैयाघ्रैः।* Ibid., Droṇa, 79.2.
18. *घोरऋक्षचर्मावृतम्।* Ibid., 142.35, 131.26.
19. *ऋक्षचर्मावनद्धाङ्गं।* Ibid., 150.11.
20. *वृताश्च द्वीपिचर्मभिः* Ibid., Udyoga, 152.6.
21. *रथाः पंचशताश्चान्ये ह्रादिनश्चर्मवर्मिणः* Ibid., Karṇa, 35.33.
22. *सिंहव्याघ्रतनुत्राणैः।* *Rām*, Sundara, 6.6.
23. *हेमपरिष्कृतम्।* Ibid., Droṇa, 2.34.
    रुक्मरथम्। Ibid., 8.8.
    सुवर्ण चित्रैः। Ibid., 79.2.
    हेमजालविचित्रेण। Ibid., Udyoga, 129.21.
    हेमजालपरिक्षिप्तैः। *Rām*, Sundara, 45.3.
    Chariots made of gold and silver were also in vogue in other countries. After the battle of Magiddo the Chariots of Cannanite rulers were captured by Thutmose III of Egypt (fifteenth century BCE). These were made of gold and silver. A. Cotterell, *Chariot*, p. 19.
24. *हेमचन्द्रकैः।* Ibid., Droṇa, 79.5.
    तारासहस्त्रखचितम्। Ibid., 122.78.
25. *किङ्किणीकिना।* Ibid., Udyoga, 129.21.
26. *वद्धध्वजपताकिनः।* Ibid., 152.9.
    सध्वजाः सपताकाश्च। Ibid., 152.4, 81.17.
27. *कांचनमालिनः।* Ibid., Virāṭa, 36.3; also Udyoga, 152.12.
28. *वैयाघ्रपरिवारणम्।* Ibid., Udyoga, 81.18; 152.6.
29. *मृगैश्चन्द्रार्धचन्द्रैश्च स रथः समलंकृतः।* *Rām*, Yuddha, 80.13.
30. *अश्वागार।* *MB*., Śānti, 69.52.
31. Ibid., Āraṇyaka, 69.10.
32. *तथैव रथशालासु प्रादुरासीद्धुताशनः।*
    ध्वजाश्चापि व्यवशीर्यन्त भरतानामभूतये।। Ibid., Sabhā, 72.22.
33. *सिंहव्याघ्रतनुत्राणैर्दान्तकाञ्चनराजतीः।*
    घोषवद्धिर्विचित्रैश्च सदा विचरितं रथैः।।
    बहुरत्नसमाकीर्णा परार्ध्यासनभूषितम्।
    महारथसमावापं महारथसमासनम्।। *Rām*, Sundara, 6.6-7.
34. *संग्रामिको रथः।* *MB*, Anuśāsana, 53.27; At the time of Rājasūya rulers presented Yudhiṣṭhira with chariots of various shapes and sizes. *रथांश्च विविधाकारान्।* Ibid., Sabhā, 51.33.
35. *MB*, Karṇa, 66.63-4.

36. J. Keegan, *A History of Warfare*, p.167.
37. *MB*, Karṇa, 24.67-76.
38. Ibid., Śānti, 236.9-12.
39. *छिन्नत्रिवेणुचक्राक्षान् हतयोधाश्वसारथीन्।*
विध्वस्तायुधतूणीरान् समुन्मथितकेतनान्।।
संछिन्नयोक्त्ररश्मीकान् वित्रिवेणुन्विकूबरान्।
विध्वस्तबन्धुरयुगान् विशस्तायुधमण्डलान्।
रथान् विशकलीकुर्वन् महाभ्राणीव मारुतः।। Ibid., Karṇa, 12.11-12.
40. *कूबरं रथचक्राणि ईषा योक्त्राणि चाभिभो।*
युगं चैवानुकर्षं च शरभूतमभूत्तदा।। Ibid., Śalya, 13.7.
ईषाणामनुकर्षाणां त्रिवेणूनां च भारत।
अक्षाणामथ योक्त्राणां प्रतोदानां च सर्वशः।। Ibid., 13.14.
छत्राणां व्यजनैः सार्धं मुकुटानां च राशयः। Ibid., 13.16
गन्धर्वनगराकारान्विधिवत्कल्पितान्रथान्।
वीषामुखान्वित्रिवेणून्व्यस्तदण्डकबन्धुरान्।।
विजङ्घकूबराक्षांश्च विनेमीननरानपि।
विचक्रोपस्करोपस्थान्भग्नोपकरणानपि।। Ibid., 35.31-2.
अपस्करैरधिष्ठानैरीषादण्डैश्च बन्धुरैः।
अक्षैर्विमथितैश्चक्रैर्भग्नैश्च वहुधा रथैः।। Ibid., Droṇa, 40.18.
युगमीषां वरूथं च तथैव ध्वजसारथी।
अश्वांस्त्रिवेणुं तल्पं च तिलशोऽभ्यहन् रथम्।। Ibid., Āraṇyaka, 231.5.
तस्यार्जुनिः शिरोग्रीवं पाणिपादं धनुर्हयान्।
छत्रं ध्वजं नियन्तारं त्रिवेणुं शम्युपस्करम्।।
चक्रे युगेषां तूणीराननुकर्षं च सायकैः।
पताकां चक्रगोप्तारौ सर्वोपकरणानि च।।
Ibid., Droṇa, 37.5-6; also 35.31-32; 43.16-17; Karṇa, 12.10-12.
अनुकर्षैरुपासङ्गैः पताकभिर्ध्वजैस्तथा।
उपस्करैरधिष्ठानैरीषादण्डकबन्धुरैः।।
चक्रैः प्रमिथतैश्चित्रैरक्षैश्च बहुधा रणे।
योगैर्योक्त्रैः कलापैश्च धनुर्भिः सायकैस्तथा।। Ibid., Droṇa, 148.42-3 (BE)
हैमत्रिकोशं रजतत्रिवेणुम्। Ibid., Karṇa, 26.56.
41. Ibid., Āraṇyaka, 230.29-31.
42. Ibid., 231.4-5; Duryodhana was accompanied by 8,000 chariots. Ibid., 228.26.
43. Monier-William, *Sanskrit-English Dictionary*, pp. 171, 300.
44. E.W. Hopkins, op. cit., p. 184.
45. Ibid., p. 185; In an encounter Bhīma is said to have destroyed the *kūvara* of Karṇa's chariot. *MB*, Droṇa, 163.14.
46. *स पीडितो महाबाहुर्गृहीत्वा रथकूबरम्।* Ibid., *Virāṭa*, 59.43.
47. *कूवरं रथचक्राणि ईषा योक्त्राणि चाभिभो।* Ibid., *Śalya*, 13.7.

48. रथनेमिनिनादैश्च वभूवाकुलिता मही। Ibid., Udyoga, 17.16 (BE).
49. Hopkins has described it in some detail. op. cit., pp. 186-7; It was perhaps the same as *apaskara*.
50. *MB*, Udyoga, 152.3; Bhīṣma, 85.34; Droṇa, 37.6, 40.18; Śalya, 13.7.
51. हैमत्रिकोषम्। Ibid., Karṇa, 26.56; The chariot of Jarāsandha called Sodaryavāna could accomodate two warriors and one charioteer. *MB*, Sabhā, 22.14.
52. E.W. Hopkins, op. cit., p. 250.
53. Ibid., pp. 185-6.
54. S.D. Singh, op. cit., p. 38.
55. रजतत्रिवेणुम्। *MB*, Karṇa, 26.56.
56. Ibid., Droṇa, 35.32.
57. E.W. Hopkins, op. cit., p. 193.
58. Ibid., p. 184.
59. मुक्तावज्रमणिस्वर्णैर्भूषितं तदयस्मयम्।
वरूथं विबभौ तस्य ज्योतिर्भिः खमिवावृतम्॥ *MB*, Droṇa, 42.5.
60. E.W. Hopkins, op. cit., pp. 179, 184.
61. सारथिं चास्य भल्लेन रथनीडादपातयत्। *MB*, Bhīṣma, 49.26; also see Karṇa, 45.16.
62. Monier-Williams, *A Sanskrit-English Dictionary*, p. 866.
63. येन कर्णो महावाहू रथोपस्थे निपातितः। *MB*, Karṇa, 35.1; Āraṇyaka, 231.5; also Droṇa, 36.33; Ram, Yuddha, 88.54.
64. Ibid., Āraṇyaka, 230.30.
65. वस्त्रैश्छत्रैश्च विध्वस्तैः। Ibid., Droṇa, 113.22.
66. E.W. Hopkins, op. cit., p. 188.
67. A warrior, if injured reclined against the staff. व्यपाश्रयत च ध्वजम्। Ibid., Droṇa, 171.54.
68. ध्वजश्च दिव्यो द्युतिमान् वानरो विस्मयंकरः। Ibid., Karṇa, 22.48.
69. केतुनापंचतारेण तालेन भरतर्षभ। Ibid., Bhīṣma, 44.48.
70. For detailed description of the *dhvajas* in the epic times, see U.P. Thapliyal, *The Dhvaja*, pp. 53-67.
71. *MB*, Droṇa, 113.22.
72. बद्धारिष्टा बद्धकक्ष्या बद्धध्वजपताकिनः। Ibid., Udyoga, 152.9.
73. ईशाबन्धं चक्रबन्धं रथबन्धं तथैव च।
प्रणाशयदमेयात्मा धृष्टद्युम्नस्य स द्विजः॥ Ibid., Droṇa, 164.132.
74. प्रदक्षिणमुपावृत्य समारोहं रथोत्तमम्। Ibid., Āraṇyaka, 164.34.
75. प्रदक्षिणमुपावृत्य समारुह्य महामनाः। Ibid., Sabhā, 42.55.
76. जैत्रैः सग्रामिकैर्मन्त्रैः पूर्वमेव रथोत्तमम्।
अभिमन्त्रितमर्चिष्मानुदयं भास्करो यथा॥ Ibid., Droṇa, 60.16.

77. *बाणबाणासनी वाहं प्रदिक्षणमवर्तत।।* Ibid., 60.14.
78. *तं रथं रथिनां श्रेष्ठः कर्णोऽभ्यर्च्य यथाविधि।*
सम्पादितं ब्रह्मविदा पूर्वमेव पुरोधसा।।
कृत्वा प्रदक्षिणं यत्नादुपस्थाय च भास्करम्। Ibid., Karṇa, 26.8-9.
79. *रथं प्रदक्षिणं कृत्वा समारोहच्चमूपतिः।* *Rām*, Yuddha, 53.7; also see 78.8, 102.17.
80. *MB*, Ādi, 212.3; Sabhā, 2.13.
81. *वाजिभिः सैन्यसुग्रीवमेघपुष्पबलाहकैः।* Ibid., Udyoga, 81.19.
82. Ibid., Āraṇyaka, 255.12; Droṇa, 81.39; The state chariot (*राज रथ*) of Yudhiṣṭhira, called *Jaitra*, was pulled by eight horses. Ibid., Sabhā, 54.5-6.
83. Ibid., Bhīṣma, 49.6.
84. Ibid., Virāṭa, 52.21.
85. Ibid., Droṇa, 142.24.
86. Ibid., Śalya, 24.20.
87. *रथांश्च विविधाकाराज्जातरूपपरिष्कृतान्।*
हयैर्विनीतैः संपन्नान्वैयाघ्रपरिवारणान्।। Ibid., Sabhā, 47.29; Also 48.36, 49.5.
88. *चतुर्युजो रथाः सर्वे।* Ibid., Udyoga, 152.10.
89. *Rām*, Yuddha, 89.39; There is also a reference to his chariot being pulled by four tigers. *Rām*, Sundara, 48.18. At another place it is said that he mounted a chariot pulled by good asses. *रथं खरश्रेष्ठसमाधि युक्तम्।* *Rām*, Yuddha, 73.8.
90. Ibid., Araṇya, 28.29.
91. Ibid., 26.8.
92. *द्रुतं सूतसमायुक्तं युक्ताष्टतुरगं रथम्।* *Rām*, Yuddha, 95.33; Only one charioteer is said to be managing these horses.
93. Ibid., Sundara, 46.28.
94. *अष्टाश्ववरैः।* Ibid., 47.4.
95. *युक्ते हयसहस्रेण विशाले स्यन्दने स्थितः।* Ibid., Yuddha, 71.12.
96. *यस्य याते न पश्यन्ति भूमौ प्राप्तं पदं पदम्।*
दक्षिणं यो धुरं युक्तः सुग्रीवसदृशो हयः।।
योऽयं धुरं धुर्यवरो वामं वहति शोभनः।
तं मन्ये मेघपुष्पस्य जवेन सदृशं हयम्।।
योऽयं काञ्चनसंनाहः पार्ष्णिं वहति शोभनः।
वामं सैन्यस्य मन्ये तं जवेन बलवत्तरम्।।
योऽयं वहति मे पार्ष्णिं दक्षिणामञ्चितोद्यतः।
बलाहकादपि मतः स जवे वीर्यवत्तरः।। *MB*, Virāṭa, 40.18-21; When Kṛṣṇa mounted his chariot to proceed to Kaurava court the horses Śaibya and Sugrīva were yoked on the right and left, Meghapuṣpa and Balāhaka followed them respectively.

दक्षिणं चावहच्छैब्यः सुग्रीवः सव्यतोऽवहत्।
पृष्टवाहौ तयोरास्तां मेघपुष्पबलाहकौ।। Ibid., Udyoga, 94, p.2315 (BE).

97. E.W. Hopkins, op. cit., p. 185.
98. S.D. Singh, *Ancient Indian Warfare*, p. 42.
99. *उष्ट्राश्वतरयुक्तानि यानानि। MB*, Anuśāsana, 119.13; Also 65.3.
100. *रासभ युक्तेन स्यन्दनेनाशुगामिना।* Ibid., Ādi, 143.7 (BE); *रथेन खरयुक्तेन। Rām*, Sundara, 44.5; *सहस्रखरसंयुक्तो रथो।* Yuddha, 69.4.
101. *एकेनापि पतत्यह्ना योजनानि चतुर्दश। MB*, Udyoga, 84.12; Anuśāsana, 66.3.
102. *स च मायामयो दिव्यः खरयुक्तः खरस्वनः।*
प्रत्यदृश्यत हेमाङ्गो रावणस्य महारथः। Rām, Araṇya, 49.19; *रथंखरश्रेष्ठ-समाधियुक्तम्।* Yuddha, 73.8.
103. *MB*, Karṇa, 27.6; also Anuśāsana, 65.3.
104. *हस्तिरथम्।* Ibid., Anuśāsana, 63.16.
105. *गजयुक्ता महाराज रथाः षड्विंशतिस्तथा।* Ibid., Sabhā, 48.29.
106. *उष्ट्राश्वतरयुक्तानि यानानि।* Ibid., Anuśāsana, 119.13; In Mausala Parva (7.33) there is a reference to chariots pulled by horses, bulls, camels and asses.
107. Ibid., Droṇa, 22; In preparing the above list Bombay edition (Geeta Press) has also been used.
108. *शैब्यस्तु शुकपत्राभः सुग्रीवः किंशुकप्रभः।*
मेघपुष्पो मेघवर्णः पाण्डुरस्तु बलाहकः।। MB, p. 2315 (BE).
109. S.D. Singh, op. cit., p. 43.
110. In the *Aśvaśāstram* one chapter is entitled *वर्ण लक्षणम् (1.31)*. Besides, mentioning *श्वेत, शोण, कृष्ण* and *हरित* types the horses are also categorised as follows:
सारोऽश्वः, कर्कोऽश्वः, कुक्कुटोऽश्वः, कौञ्चपत्रोऽश्वः, सिन्धुवारकोऽश्वः, सिंहोऽश्वः, शुक्लरोमकोऽश्वः, कालरोमकोऽश्वः, मृदुरोमकोऽश्वः, कालाक्षिकोऽश्वः, श्वेताक्षिकोऽश्वः, पाटलोऽश्वः, रक्तपाटलोऽश्वः, चक्रवाकोऽश्वः, सिद्धार्थकोऽश्वः, कपिलोऽश्वः, धूर्मवर्णोऽश्वः, खञ्जरीट्यऽश्वः, करण्टकोऽश्वः, कुर्वरोऽश्वः, etc.
The colours of the horses of some gods are stated as follows:
सर्वश्वेतो शशिनो
हरयः कनकप्रभाः सुरेन्द्रस्य।
कृष्णा यमस्य वाहा
कर्काः शुभलक्षणा विष्णो।।
शुकपत्रनिभा भानोः
किंशुकवर्णा हया हुताशस्य।
असिताभ्रनिभा वरुणस्य
कुवेरस्य तु कुमुदनिभास्तुरङ्गमाः।। Aśvaśāstram, 1.29-30.
The colours of the horses of kings, Brāhmaṇas, Kṣatriyas, officials,

etc., are also mentioned in Rājavāhādhyāyaḥ (*Aśvaśāstram*, 100-111).

111. *दंशितैर्हरिभिर्युक्तं रथमास्थाय कांचनम्।* *MB*, Āraṇyaka, 18.1; *चामरापीडकवचाः।* Ibid., Droṇa, 18.31.
112. *अश्वाः सैन्धवा हेममालिनः।* Ibid, Droṇa, 8.15.
113. *जाम्बूनदैर्जालैः संछन्नान्।* Ibid., 106.30.
114. *स्वर्णजाल परिच्छदाः।* Ibid., 22.41.
115. *रुक्मपत्रैरुरच्छदैः।* Ibid., 22.29.
116. *चित्रकांचनसंनाहैर्वाजिमुख्यैर्विशाम्पते।* Ibid., 122.80.
117. *चित्रकांचनसनाहैः सदश्वैर्वेगवत्तरैः।* Ibid., 122.83.
118. *निकृत्तवर्मकवचान्।* Ibid.; also, 35.39.
119. *वैयाघ्रैर्हेमचन्द्रकैः।* Ibid., 79.5.
120. *वैडूर्यजालसंछन्नैः।* Ibid., 23.73 (BE).
121. Ibid., Droṇa, 22.29, 22.41, 106.30.
122. Ibid., 79.5.
123. Ibid., Karṇa, 14.43.
124. Ibid., Droṇa, 79.5.
125. *हेमभाण्डपरिच्छदैः।* Droṇa, 122.79; also see E.W. Hopkins, op. cit., pp. 257-8.
126. A. Staal, *Discovering the Vedas*, p. 17.
127. *बनायुजान् पार्वतीयान् काम्बोजारट्टबाह्लिकान्।* *MB*, Droṇa, 35.36, 97.26. Kāmboja people shaved their heads. Ibid., 95.20. The dowry presented by Kṛṣṇa to Pāṇḍavas at the time of Subhadrā's marriage included one lakh horses of Bāhlīka country (Ādi, 220.51 BE) and one thousand she-mules (Ādi, 220.48).
आजानेयैर्महावेगैर्नानादेशसमुत्थितैः।
पर्वतीयैर्नदीजैश्च सैन्धवैश्च हयोत्तमैः।। Ibid., 79.8. Ājāneya horses pulled the chariot of Duryodhana. Ibid., Śānti, 124.10.
ततः काम्बोजमुख्यानां नदीजानां च वाजिनाम्।
आरट्टानां महीजानां सिन्धुजानां च सर्वशः।।
वनायुजानां शुभ्राणां तथा पर्वतवासिनाम्।
ये चापरे तित्तिरिजा जवना वातरंहसः।।
सुवर्णालंकृतैरेतैर्वर्मवखिः सुकल्पितैः।। Ibid., Bhīṣma, 86.3-5.
कम्बोजविषये जातैर्बाह्लीकैश्च हयोत्तमैः।
वनायुजैनदीजैश्च पूर्णा हरिहयोत्तमैः।। *Rām*, Bāla, 6.22.
कौलूतका हयाश्चित्रा। *MB*, Droṇa, 79.7.
The army of Śakuni included many horses of foreign breed:
आजानेयैः सैन्धवैः पार्वतीयैर्नदीजकाम्बोज वनायुबाह्लिकैः। *MB*, Karṇa, 4.96.
Sātyaki killed many such horses in an encounter:
वनायुजान् पर्वतीयान् काम्बोजारट्ट बाह्लिकान्।
तथा हयवरान् राजन् निजघ्ने तत्र सात्यकिः।। Ibid., Droṇa, 97.26

During the consecration ceremony of Yudhiṣṭhira the Bāhlīka king brought a chariot, yoked with white Kāmboja horses. Sabhā, 49.5; Also 52.20. The horses of Bāhlīka country were highly valued. *बाह्लीकजातैर्हयोत्तमैः।* Ibid., Udyoga, 84.6. The Kāmboja horses were equally prized is also evident from the fact that Karṇa announced an award of a golden chariot yoked with Kāmboja horses to any one who informed him about the location of Arjuna. *MB*, Karṇa, 27.8; In the *Amarakośa* (2.8.45), horses have been classified in four categories, viz., Vanāyuja, Pārasīka, Kāmboja and Bāhlīka. These all appear to be imported types. Ājāneya has been called the top breed horse by *Amara* (2.8.44). In the *Raghuvaṁśa* the chariot of prince Aja is said to have been pulled by the horses from Vanāyu country. *वनायुदेश्याः।* 5.73.

128. Ibid., Droṇa, 22.16-18, 97.26; The chariot of Droṇa was pulled by horses from Sindh (*सैन्धवाः*). Ibid., Droṇa, 8.16.
129. Ibid., Ādi, 158.45; At the time of Rājasūya he presented 400 horses to Yudhiṣṭhira. Ibid., Sabhā, 48.22.
130. Ibid., Āraṇyaka, 253.6.
131. In *Aśvaśāstra, Kulalaksaṇa*, 1-64, the attributes of these horses have also been described.
132. *यवना हयानुपादाय पर्वतीयान् मनोजवान्।* Sabhā, 47.13.
133. *अजाविकं गोहिरण्यं खरोष्ट्रं फलजं मधु।* Ibid., 47.11.
134. *आजानेयान् हयाञ्शीघ्रानादायानिलरंहसः।* Ibid., 47.13; Ājāneya horses commanded great respect and the chariot of Duryodhana was pulled by these horses. Ibid., 45.9.
135. *अश्वान् मनोजवान्।* Ibid., 47.18.
136. *कृष्णग्रीवान् महाकायान् रासभाञ्शतपातिनः।*
आहार्षुर्दशसाहस्रान् विनीतान् दिक्षु विश्रुतान्।। Ibid., 47.21. The countries, which made these presents were indeed located beyond the western borders of India. Ibid., 47.19.
137. *महागमान्दूरगमान्गणितानर्बुदं हयान्।* Ibid., 47.26.
138. Pierre Ponafidine quoted by Ruchira Banerjee, *Indica*, September 2015, vol. 52.
139. *रासभयुक्तेन स्यन्दनेनाशुगामिना।* Ibid., Ādi, 143.7 (BE).
140. *एकेनापि पतत्यह्ना योजनानि चतुर्दश।*
यानमश्वतरीयुक्तं दास्ये तस्मै तदप्यहम्।। Ibid., Udyoga, 84.12.
141. E. Mireaux, *Daily life in the Time of Homer*, p. 73.
142. *उपास्थितमसंहार्यैर्हयैः स्यन्दनयायिभिः।* Rām, Sundara, 6.5.
143. *हयैर्महाजवैर्युक्तं।* Ibid., Yuddha, 57.25.
144. *आरूतारते हयारतत्र रथमूहुरविजलाजा.।*
मण्डलान्यभिधावन्ति तदद्भुतमिवाभवत्।। Ibid., 90.28-9.

145. सव्यं चक्रं युधामन्युरुत्तमौजाश्च दक्षिणम्। *MB*, Bhīṣma, 95.21; Śalya, 15.24.
146. चन्द्रसूर्यप्रकाशाभ्यां चक्राभ्यां समलंकृतम्। Ibid., Udyoga, 81.16.
147. C.V. Vaidya, *Epic India*, p. 257.
148. S.D. Singh, op. cit., p. 42.
149. अष्टचक्रसमायुक्तम्। *MB*, Droṇa, 142.34.
150. Ibid., 150.12, 131.28.
151. सचक्राश्च विचक्राश्च रथैरेव महारथाः। Ibid., 19.55.
152. चक्राणि विविधानि च। Ibid., Karṇa, 56.38.
153. E.W. Hopkins, op. cit., p. 181.
154. रथनेमिनिकृत्ताश्च। *MB*, Bhīṣma, 112.128.
155. Ibid., Virāṭa, 32.30; also see Bhīṣma, 50.69; Droṇa, 66.35; Karṇa, 11.31, 34.44.
156. Ibid., Bhīṣma, 50.69.
157. Ibid., Ādi, 125.21.
158. युधिष्ठिरो रथश्रेष्ठः सर्वत्र तु धनंजयः।
प्रथितः सागरान्तायां रथयूथपयूथपः।। *MB*, Ādi, 123.41-2.
159. रथयुद्ध। Ibid., 138.4 (BE).
160. *MB*, Udyoga, 162-9; In Āraṇyaka Parva (84.7) Yudhiṣṭhira calls Bhīṣma and Droṇa as *Atirathins*. In Ādi Parva Nakula and Sahadeva have been called *Atirathins* (138.19).
161. *MB*, Udyoga, 165.6; In the *Rāmāyaṇa* Meghanāda has been ranked as *Atirathin* (Yuddha, 80.33); In another context, among the sons of Dhṛtarāṣṭra, Duryodhana, Duḥśāsana, Duḥsaha, Durmarṣeṇa, Vikarṇa, Citrasena, Vivinśati, Jaya, Satyavrata, Purumitra and Yuyutsu are called *mahārathins*. MB, Ādi, 63.119-20 (BE).
162. *MB*., Sabhā, 13.56-59.
163. एकैको यत्र लभते सहस्रपरमां भृतिम्। Ibid., 54.20.
164. शुक्लवासाः सितोष्णीषः सर्वशुक्लविभूषणः। Ibid., Udyoga, 179.14.
165. दंशितः पाण्डुरेणाहं कवचेन वपुस्मता। Ibid., 179.12.
166. ध्वजेन च महाबाहो सोमालंकृतलक्ष्मणा।
धनुर्धरो वद्धतूणो वद्धगोधाङ्गुलित्रवान।। Ibid., 180.8.
167. कवची सतलत्राणो वद्धगोधाङ्गुलित्रवान्। Ibid., Āraṇyaka, 38.16.
168. तस्य कृष्णायसं वर्म हेमचित्रम्। Ibid., Droṇa, 102.55.
कवचैश्च हिरण्मयैः। Ibid., 73.25.
स कांचनविचित्रेण कवचेन समावृतः। Ibid., 65.6.
169. कांस्ये निर्भिद्य वर्मणी। Ibid., 150.24.
170. Ibid., Āraṇyaka, 161.16; Virāṭa, 55.24.
171. सोष्णीषं सशिरस्त्राणम् क्षुरप्रणान्वपातयत्। *MB*, Karṇa, 38.28.
172. शिरस्त्राणम् चार्कसमानभासम्। Ibid., Droṇa, 2.23.

173. *जाम्बूनदशिरस्त्राणः।* Ibid., 65.6.
174. *उष्णीषाणि नियच्छन्तः पुण्डरीकनिभैः करैः।*
अन्तरीयोत्तरीयाणि भूषणानि च सर्वशः।। Ibid., Udyoga, 150.20.
175. Ibid., Bhīṣma, 23.12-19.
176. *दध्मौ शंखं महावेगमार्षभेणाथ माधवः।* Ibid., Droṇa, 122.42.
177. Ibid., Bhīṣma, 23.14.
178. Ibid., Bhīṣma, 23.15-16.
179. *अश्वांश्च दिव्यानिच्छेयं पाण्डुरान् वातरंहसः।*
रथं च मेघनिर्घोषं सूर्यप्रतिमतेजसम्।। Ibid., Ādi, 223.18 (BE).
ततः श्वेतैर्हयैर्युक्ते महति स्यन्दने स्थितौ। Ibid., Bhīṣma, 23.14 (BE).
180. *रथं च दिव्याश्वयुजं कपिप्रवरकेतनम्।*
उपेतं रजतैरश्वैर्गान्धर्वैर्हेममालिभिः।। Ibid., Ādi, 224.10 (BE).
पाण्डुराभ्रप्रतीकाशैर्मनोवायुसमैर्जवे।
सर्वोपकरणैर्युक्तमजय्यं देवदानवैः।। Ibid., Ādi, 224.11 (BE).
भानुमन्तं महाघोषं सर्वरत्नमनोरमम्।
संसर्ज यं सुतपसा भौमनो भुवनप्रभुः।। Ibid., Ādi, 224.12 (BE).
प्रजापतिनिर्देश्यं यस्यं रूपं रवेरिव।
यं स्म सोमः समारुह्य दानवानजयत् प्रभुः।। Ibid., Ādi, 224.13 (BE).
ब्रह्मेशानेन्द्रवरुणानवहद् यः पुरा रथः। Ibid., Droṇa, 127.1. (BE).
तापनीया सुरुचिरा ध्वजयष्टिरनुत्तमा।
तस्यां च वानरो दिव्यः सिंहशार्दूलकेतनः।। Ibid., Ādi, 224.15 (BE).
नादेन रिपुसैन्यानां येषां संज्ञा प्रणश्यति।
स तं नानापताकाभिः शोभितं रथसत्तमम्।। Ibid., Ādi, 224.17.
There are many more references to the chariot of Arjuna in the *Mahābhārata*. See Virāṭa, 2.16-19, 61.1-5; Udyoga, 55.3-12; Bhīṣma, 22.9, 23.14, 52.18-19.
181. On the request of Arjuna Kṛṣṇa agreed to become his charioteer.
उपपन्नमिदं पार्थ यत्स्पर्धेथा मया सह।
सारथ्यं ते करिष्यामि कामः सम्पद्यतां तव।। Ibid., Udyoga, 7.35.
182. *युक्तं परमकाम्बोजैस्तुरगैर्हेममालिभिः।*
आदित्योदयवर्णस्य धुरं रथवरस्य तु।।
दक्षिणामवहत्सैन्यः सुग्रीवः सव्यतोऽवहत्।
पार्ष्णिवाहो तु तस्यास्तां मेघपुष्पबलाहकौ।। Ibid., Sauptika, 13.1-2.
183. *शैव्यस्तु शुकपत्राभः सुग्रीवः किंशुकप्रभः।*
मेघपुष्पो मेघवर्णः पाण्डुरस्तु बलाहकः।। Ibid., Udyoga, 94 (GE).
184. *रथ आरोप्यतां शङ्खश्चक्रं च गदया सह।*
उपासंगाश्च शक्त्यश्च सर्वप्रहरणानि च।। Ibid., Udyoga, 81.12.
तं दीप्तमिव कालाग्निमाकाशगमिवाध्वगम्।
चन्द्रसूर्यप्रकाशाभ्यां चक्राभ्यां समलंकृतम्।।
अर्धचन्द्रैश्च चन्द्रैश्च मत्स्यैः सामृगपक्षिभिः।

पुष्पैश्च विविधैश्चित्रं मणिरत्नैश्च सर्वशः।।
मणिहेमविचित्राङ्गं सुध्वजं सुपताकिनम्।।
सुपस्करमनाधृष्यं वैयाघ्रपरिवारणम्। Ibid., 81.15-18.

185. सुघोषः पतगेन्द्रेण ध्वजेन युयुजे रथः।। Ibid., 81.20.
186. ततो रथेन शुभ्रेण महता किङ्किणीकिना।
हयोत्तमयुजाः शीघ्रमुपातिष्ठत दारुकः।। Ibid., 94.12.
There are many more descriptions of Kṛṣṇa's chariot in the *Mahābhārata*.
187. संह्लादनो राजरथो य इहास्मानुपावहत्।
जैत्रो रथवरः पुण्यो मेघसागरनिःस्वनः।।
अष्टौ यं कुररच्छायाः सदश्वा राष्ट्रसंमताः।
वहन्ति नैषा मुच्येत पदा भूमिमुपस्पृशन्।। Ibid., Sabhā, 54.5-6.
188. अयं सहस्रसमितो वैयाघ्रः सुप्रवर्तितः।
सुचक्रोपस्करः श्रीमान् किङ्किणीजालमण्डितः।। Ibid., 54.4
189. एतद् राजन् धनं मह्यं तेन दीव्याम्यहं त्वया। Ibid., 54.6.
190. एवमुक्त्वा महाराज तं रथं हेमभूषितम्।
दन्तवर्णैर्हयैर्युक्तं कालवालैमहारथः।। *MB*, Karṇa, 69.26
191. ततश्चिन्तयमानस्य गुडाकेशस्य धीमतः।
रथो मातलिसंयुक्त आजगाम महाप्रभः।।
नभो वितिमिरं कुर्वञ्जलदान् पाटयन्निव।
दिशः सम्पूरयन् नादैर्महामेघरवोपमैः।।
असयः शक्तयो भीमा गदाश्चोग्रप्रदर्शनाः
दिव्यप्रभावा प्रासाश्च विद्युतश्च महाप्रभाः
तथैवाशनयस्तत्र चक्रयुक्ताहुडागुडाः।
वायुस्फोटाः सनिर्धाता बर्हिमेघनिभस्वनाः।।
दशवाजिसहस्राणि हरीणां वातरंहसाम्।
वहन्ति यं नेत्रमुषं दिव्यं मायामयं रथम्।। Ibid., Āraṇyaka, 43.2-7.
192. रथेनानेन मघवा जितवान् शम्बरं युधि।
निमुचिं बलवृत्रौ च प्रह्लादनरकावपि।। Ibid., 165.18.
193. Ibid., 165-9.
194. Ibid., 274; also see *Rām*, Yuddha, 102.9-12.
195. यः स सोदर्यवान् नाम द्वियोधः कृष्णसारथिः।
अभ्यासघाती संदृश्यो दुर्जयः सर्वराजभिः।। Ibid., Sabhā, 22.14.
196. शुशुभे रथवर्योऽसौ दुर्जयः सर्वधन्विभिः।। Ibid., 22.15.
197. तप्तचामीकराभेण किङ्किणीजालमालिना।
मेघनिर्घोषनादेन जैत्रेणामित्रघातिना।। Ibid., 22.17.
198. हयैर्दिव्यैः समायुक्तो रथो वायुसमे जवे। Ibid., 22.20.
असङ्गी देवविहितस्तस्मिन् रथवरे ध्वजः।
योजनाद् ददृशे श्रीमानिन्द्रायुधसमप्रभः।। Ibid., 22.21.
199. येन शक्रो दानवानां जघान नवतीर्नव। Ibid., 22.18.

200. *यं लेभे वासवाद्राजा वसुस्तस्माद् बृहद्रथः।*
वृहद्रथात् क्रमेणैव प्राप्तो बार्हद्रथं नृपम्।। Ibid., 22.27.
201. *तेनैव रथमुख्येन तरुणादित्यवर्चसा।*
धर्मराजविसृष्टेन दिव्येनानादयन् दिशः।। Ibid., 22.54.
202. *रथमास्थाय रुचिरं राजतं पाण्डुरैर्हयैः।*
सुपस्करं स्वधिष्ठानं वैयाघ्रपरिवारणम्।।
उपपन्नं महाशस्त्रैः सर्वोपकरणान्वितम्।
तत्कुलीनेन वीरेण हयशास्त्रविदा नृप।।
युक्तं सूतेन शिष्टेन बहुशो दृष्टकर्मणा। Ibid., Udyoga, 179.10-12.
203. *ते हयाश्चोदितास्तेन सूतेन परमाहवे।*
अवहन् मां भृशं राजन् मनोमारुतरंहसः।। Ibid., 179.16.
204. *यस्तु नीलानुसारेण पञ्चतारेण केतुना।* Ibid., Virāṭa, 50.18.
205. *यस्य तारार्कचित्रोऽसौ रथे ध्वजवरः स्थितः।* Ibid., 50.19.
206. *केतुना पञ्चतारेण तालेन भरतर्षभ।* Ibid., Bhīṣma, 44.48.
207. *खचरनगरकल्पं कल्पितं शास्त्रदृष्ट्या*
चलदनिलपताकं ह्लादिनं वल्गिताश्वम्।
स्फटिकविमलकेतुं तापनं शात्रबाणां
रथवरमधिरूढः संजहारारिसेनाम्।। Ibid., Droṇa, 7.43.
208. *तस्य शोणितदिग्धाङ्गाः शोणास्ते वातरंहसः।*
आजानेया हया राजन्नविभ्रान्ता श्रियं दधुः।। Ibid., 7.10.
209. *रुक्मरथं वैयाघ्रपरिवारणम्।* Ibid., 8.8; also Bhīṣma, 20.11.
210. *अतीव शुशुभे तस्य ध्वजः कृष्णाजिनोत्तरः।*
कमण्डलुर्महाराज जातरूपमयः शुभः।। Ibid., 23.82 (BE).
पताकिनं शोणहयं वेदीकृष्णाजिनध्वजम्। Ibid., 63.30; also Bhīṣma, 17.24.
211. *रथं चाग्र्यं हेमजालावनद्धं*
रत्नैश्चित्रं चन्द्रसूर्यप्रकाशैः।
द्रव्यैर्युक्तं सम्प्रहारोपपन्नै
र्वाहैर्युक्तं तूर्णमावर्तयस्व।। *MB*, Droṇa, 2.27; also Karṇa, 26.56.
अश्वानग्र्यान् पाण्डुराभ्रप्रकाशान्
पुष्टान् स्नातान् मंत्रपूताभिरद्भिः। Ibid., 2.26.
वैयाघ्रचर्माणम्। Ibid., Karṇa, 26.56.
212. *मेघस्वनेनार्क इवामितौजाः।।* Ibid., 2.36.
213. *अश्वांश्च चतुरः श्वेतान्।* Ibid., 122.61.
हसवर्णान् हयान्। Ibid., Karṇa, 34.3.
214. *तप्तैर्भाण्डैः काञ्चनैरभ्युपेता-*
ञ्शीघ्राञ्शीघ्रं सूतपुत्रानयस्व। Ibid., Droṇa, 2.26.
215. *एतां रौक्मीं नागकक्ष्यां च जैत्रीं।*
जैत्रं च मे ध्वजमिन्दीवराभम्।। Ibid., 2.25.
216. *सुवर्णमुक्तामणिवज्रशालिना।* Ibid, 2.36.
217. *ऋश्यवर्णैर्हयैः।* Ibid., 22.2; also 107.26.

218. *विशोकेनाभिसंयत्ता मनोमारुतरंहसः।* Ibid., 102.65.
219. *जम्बुनदमयः केतुः केसरी सः नरसत्तम।* Ibid, Bhīṣma, 45.33; also 91.70.
220. *ध्वजं च भीमसेनस्य वैदूर्यमणिलोचनम्।*
भ्राजमानं महासिहं राजन्तं हष्टवानहम्॥ Ibid., Droṇa, 23.83 (BE).
221. *किंकिणीशतनिर्घोषं रक्तध्वजपताकिनम्॥*
ऋक्षचर्मावनद्धाङ्गं नल्वमात्रं महारथम्।
सर्वायुधवरोपेतमास्थितो ध्वजशालिनम्॥
अष्टचक्रसमायुक्तं मेघगंभीरनिःस्वनम्। Ibid., 23.11-13 (BE).
222. *मत्तमातङ्गसंकाशा लोहिताक्षा विभीषणाः॥*
कामवर्णजवा युक्ता बलवन्तः शतं हयाः। Ibid., 23.13-14 (BE).
223. *दिवःस्पक् सुमहान् केतुः स्यन्दनेऽस्य समुच्छ्रितः॥*
रक्तोत्तमाङ्गः क्रव्यादो गृधः परमभीषणः। Ibid., 23.17-18 (BE).
224. *राक्षसोऽस्य विरूपाक्षः सूतो दीप्तास्यकुण्डलः॥* Ibid., 23.15 (BE).
225. *आयसैः काञ्चनैश्चापि पट्टैर्नद्धं सकूबरम्।*
तारासहस्रखचितं सिंहध्वजपताकिनम्॥
अश्वैर्वातजवैर्युक्तं हेमभाण्डपरिच्छदैः।
पाण्डुरैरिन्दुसंकाशैः सर्वशब्दातिगैर्दृढैः॥
चित्रकाञ्चनसंनाहैर्वाजिमुख्यैर्विशाम्पते।
घण्टाजालाकुलरवं शक्तितोमरविद्युतम्॥
वृतं सांग्रामिकैर्द्रव्यैर्बहुशस्त्रपरिच्छदम्।
रथं सम्पादयामास मेघगम्भीरनिस्वनम्॥ Ibid., 122.78-81.
226. *सिंहध्वजपताकिनम्।* Ibid., 122.78.
227. *दारुकस्यानुजस्तूर्णं कल्पनाविधिकल्पितम्।* Ibid., 122.77.
228. *एवमुक्त्वा रौक्मिणेयो यादवान् भरतर्षभ।*
दंशितैर्हरिभिर्युक्तं रथमास्थाय काञ्चनम्॥
उच्छ्रित्य मकरं केतुं व्यात्ताननमलंकृतम्।
उत्पतद्भिरिवाकाशं तैर्हयैरन्वयात् परान्॥ Ibid., Āraṇyaka, 18.1-2.
229. *तं तथा मोहितं दृष्ट्वा सारथिर्जवनैर्हयैः।*
रणादपाहरत्तूर्णं शिक्षितो दारुकिस्ततः॥ Ibid., 19.3.
230. *कल्प्यतां मे रथः शीघ्रम्।* *Rām*, Yuddha, 95.21.
231. *दिव्यास्त्रवरसम्पन्नं नानालंकारभूषितम्।*
नानायुधसमाकीर्णं किङ्किणीजालसंयुतम्॥
नानारत्नपरिक्षिप्तं रत्नस्तम्भैर्विराजितम्।
जाम्बूनदमयैश्चैव सहस्रकलशैर्वृतम्॥ Ibid., 95.30-31; also see 106.1-8.
232. *जलदस्वनं।* Ibid., 103.31.
233. *द्रुतं सूतसमायुक्तं युक्ताष्टतुरगं रथम्।*
आरुरोह तदा भीमं दीप्यमानं स्वतेजसा॥ Ibid., 95.33.
234. *ध्वजं मनुष्यशीर्षम्।* Ibid., 100.14.
235. *जाज्वल्यमानो वपुषा तपनीयपरिच्छदः।*
मृगैश्चन्द्रार्धचन्द्रैश्च स रथः समलंकृतः॥ Ibid., 80.13.

236. *जाम्बूनदमहाकम्बुर्दीप्तपावकसंनिभः।*
बभूवेन्द्रजितः केतुर्वैदूर्यसमलंकृतः।। Ibid., 80.14.
237. *अथ राक्षससिंहस्य कृष्णान् कनकभूषणान्।*
शरैश्चतुर्भिः सौमित्रिर्विव्याध चतुरो हयान्।। Ibid., 89.39; Also see 88.2.
238. *स रथं भूषयित्वाथ रुचिरं हेमभूषितम्।*
प्रासासिशरसंयुक्तं युक्तं परमवाजिभिः।।
अधिष्ठितं हयज्ञेन सूतेनाप्तोपदेशिना।
आरुरोह महातेजा रावणिः समितिंजयः।। Ibid., 90.9-10.
239. *युक्ते हयसहस्रेण विशाले स्यन्दने स्थितः।* Ibid., 71.12.
240. *चतुःसादिसमायुक्तो मेघस्तनितनिःस्वनः।।* Ibid., 71.19.
241. Ibid., 71.20-1.
242. *ध्वजशृङ्गप्रतिष्ठेन राहुणाभिविराजते।* Ibid., 71.17.
243. *रथांश्च दान्तान् सौवर्णैः शुभ्रैः पटैरलंकृतान्।* *MB*, Adi, 198.17.
244. *रथानां काञ्चनाङ्गानां किङ्किणीजालमालिनाम्।*
चतुर्युजामुपेतानां सुतैः कुशलशिक्षितैः।। Ibid., 220.45.
The dowery presented to Devakī by her father included 1,800 chariots.
245. *रथांश्च विविधाकाराञ्जातरूपपरिष्कृतान्।*
हयैर्विनीतैः सम्पन्नान् वैयाघ्रपरिवारणान्।। Ibid., Sabhā, 47.29. During Rājasūya the ruler of Sauvīra presented 300 chariots drawn by elephants to Yudhiṣṭhira. These chariots were decorated with diamonds, pearls and gold.
246. *एकः वर्णैः सुकृष्णाङ्गैर्बाह्लिजातैर्हयोतमैः।*
चतुर्युक्तान् रथस्तस्मै रौक्मान् दास्यामि षोडश।। Ibid., Udyoga, 84.6.
247. *दासीशतं चापि रथांश्च विंशतिम्।* Ibid., Karṇa, 76.40 (BE).
248. Ibid., Anuśāsana, 53.29, 63.16.
249. *प्रयाद् गर्दभयुक्तेन रथेनेहाशुगामिना।* Ibid., 28.8.
250. Ibid., 53.29.
251. *एकैको यत्र लभते सहस्रपरमां भृतिम्।*
युध्यतोऽयुध्यतो वापि वेतनं मासकालिकम्।। Ibid., Sabhā, 54.20.
252. *त्रिंशदेनं सहस्राणि रथाः काञ्चनमालिनः।* Ibid., Virāṭa, 65.12.
253. *नागेष्वश्वेषु बहुशो रथेषु च परीक्षितम्।*
परीक्ष्य च यथान्यायं वेतनेनोपपादितम्।। Ibid., Droṇa, 89.6.
254. *एतस्य रथवंशो हि तिग्मवेगप्रहारिणाम्।* Ibid., Udyoga, 163.3
255. *नीलो माहिष्मतीवासी नीलवर्मधरस्तव।*
रथवंशेन शत्रूणां कदनं वै करिष्यति।। Ibid., 163.4.
256. *अष्टौ रथसहस्राणि।* Ibid., Virāṭa, 30.28.
257. *षष्ट्या रथसहस्रैस्तु।* Ibid., Bhīṣma, 17.32.
258. *रथानामयुतैस्त्रिभिः।* Ibid., Droṇa, 16.19.
259. *रथानामयुतेनैव सोऽशपद्भ्रातृभिः सह।।*
नागाजनपदेभ्यश्च रथानामयुतं पुनः। Ibid., 16.20-21.

260. *शतं रथ सहस्राणाम्।* Ibid., Bhīṣma, 17.30.

261. *रथानामयुतं पक्षौ शिरश्च नियुतं तथा।*
पृष्टमर्बुदमेवासीत् सहस्राणि च विंशतिः।।
ग्रीवायां नियुतं चापि सहस्राणि च सप्ततिः
पक्षकोटिप्रपक्षेषु पक्षान्तेषु च बारणाः।। Ibid., 46.52-3.

262. Ibid., Ādi, 2.9, etc.; The *Amarakośa* also gives almost similar composition of army formations. It, however, adds an additional formation called *daśānīkinī* between *anīkinī* and *akṣauhiṇī*.
एकेभैकरथा त्र्यश्वा पत्तिः पञ्चपदातिका।
पत्त्यङ्गैस्त्रिगुणैः सर्वैः क्रमादाख्या यथोत्तरम्।। *Amarakośa*, 2.8.80.
सेनामुखं गुल्मगणौ वाहिनी पृतना चमूः।
अनीकिनी दशानीकिन्यक्षौहिण्यथ संपदि। Ibid. 2.8.81.
The number of chariots in an *akṣauhiṇī* (21,870) is described in *Sadāśva Dhanurveda* as follows:
खं तथा स्वरवस्विन्दुनेत्रैरक्षौहिणी मता।
अक्षौहिन्यां संप्रदिष्टा रथानां वर्मधारिणाम्।। *SD*, 166.
(The armoured chariots should number 21,870 (By computing the legends) *kha* (sky or 0) *svara* (tunes 7) Vasu (8) Indu (moon 1) *netra* (eye 2). As per the law of counting in Sanskrit *(अङ्कना वामतो गतिं;)* the digits are counted from right to left, hence the figure 21,870. It further goes on to say that a
महाक्षौहिणी comprised 1,37,00,000 chariots.
महाक्षौहिणिकां प्राहुरिमां तत्वविदो जनाः।
महाक्षौहिणिकां तु रथाः कोटिमिताः स्मृताः।
सप्तत्रिंशच्च लक्षाणि गीयन्ते तत्वभेदिभिः।। *SD*, 170.
The logic behind these exaggerated figures is not understandable.

263. E.W. Hopkins, op. cit., p. 140.

264. *एकादश सहस्राणि रथानां भरतर्षभ।*
दश दन्तिसहस्राणि सप्त चैव शतानि च।।
पूर्णे शतसहस्रे द्वे हयानां भरतर्षभ।
नरकोट्यस्तथा तिस्रो वलमेतत्तवाभवत्।।
रथानां षट्सहस्राणि षटसहस्राश्च कुञ्जराः।
दश चाश्वसहस्राणि पत्तिकोटी च भारत।।
एतद्बलं पाण्डवानामभवच्छेषमाहवे।
एत एव समाजग्मुर्युद्धाय भरतर्षभ।। *MB*, Śalya, 7.37-40.

265. *अथानयन् बलाध्यक्षाश्चत्वारो रावणाज्ञया।*
रथानां नियुतं साग्रं नागानां नियुतत्रयम्।।
अश्वानां षष्टिकोट्यस्तु खरोष्ट्राणां तथैव च।
पदातयस्त्वसंख्याता जग्मुस्ते राजशासनात्।। *Rām*, Yuddha, 95.27-28. At another place the army of Rāvaṇa is said to have comprised

10 thousand elephants, 10 thousand chariots, 20 thousand horses and more than 1 crore foot soldiers. Yuddha, 37.16.

266. *अनीकं दशसाहस्रं रथानां वातरंहसाम्।* Ibid., 93.31.

267. *रथकुंजरवाजिनाम्।* Ibid., Sundara, 31.2.

268. *तामश्वरथसम्पूर्णां रथध्वजविभूषिताम्।* Ibid., Ayodhyā, 96.13.

269. *महद्भ्यां रथवृन्दाभ्याम्। MB*, Droṇa, 156.53 (BE).

270. *रथसहस्रेण।* Ibid., 64.9.

271. *बलानि सुसमृद्धानि साश्वनागरथानि च। Rām*, Sundara, 48.8.

272. *गजानां रथिनो मध्ये रथानामनु सादिनः।*
सादिनामन्तरे स्थाप्यं पदातमिह दंशितम्।।
य एवं व्यूहते राजा स नित्यं जयते द्विषः।
तस्मादेवं विधातव्यं नित्यमेव युधिष्ठिरः।। MB, Śānti, 100.9-10.

273. *तथा कवचिना शूराः शस्त्रेषु कृतनिश्रमाः।।*
कुलीना हययोनिज्ञाः सारथ्ये विनिवेशिताः। Ibid., Udyoga, 152.8-9.

274. *सूतैः कुशलसंमतैः।* Ibid., Ādi, 213.41.

275. *हयतत्वज्ञ।* Ibid., 168.37 (BE); Virāṭa, 40.17.

276. *हयशास्त्र विदा नृप।* Ibid., Udyoga, 179.11; Obviously, literature had come to stay on the subject by now.

277. Ibid., Āraṇyaka, 19.3.

278. Ibid., Virāṭa, 35.13.

279. During a dialogue with Karṇa, Śalya spoke in detail about the qualities of an able charioteer (Karṇa, 28.5-8). After a fierce battle against Jayadratha the horses of Arjuna's chariot got injured and tired. Kṛṣṇa, as an expert charioteer, unyoked them, attended to their injuries, rested them and fed them with water. The horses thus refreshed were again yoked to the chariot. (Ibid., Droṇa, 13.16).

280. *तत्कुलीनेन वीरेण हयशास्त्रविदा नृप।*
युक्तं सूतेन शिष्टेन बहुशो दृष्टकर्मणा।। Ibid., Udyoga, 179.11.

281. *शिक्षितो ह्यस्मि सारथ्ये तीर्थतः पुरुषर्षभः।* Ibid., Virāṭa, 40.16.
दारुको वासुदेवस्य यथा शक्रस्य मातलिः।
तथा मां विद्धि सारथ्ये शिक्षितं नरपुंगव।। Ibid., 40.17
In the *Matsya Purāṇa* (215.20-1) the qualities expected of a charioeer are summed up as follows:
निगितशकुन-ज्ञानी हयशिक्षा विशारदः।
हयायु र्वेदतत्वज्ञो भुर्वा-भागविचक्षणः।।
जलाबलज्ञो रथिनः स्थिरदृष्टिः प्रियंवदः।
शुरश्च कृतविद्यश्च सारथिः परिकीर्तितः।।

282. *रथ शिक्षा विशारदो।* Ibid., Droṇa, 38.29.

283. Ibid., Udyoga, 8.29; also Karṇa, 23.4-5.

284. *Rām*, Yuddha, 89.43.
285. E.W. Hopkins, op. cit., p. 181.
286. *धूर्ययोर्हययोरेकस्तथान्यौ पार्ष्णिसारथी।*
तौ चापि रथिनां श्रेष्ठौ रथी च हयवित्तथा।। *MB*, Udyoga, 152.11; Also Virāṭa, 32.8; 45.20-23; Droṇa, 47.32; Karṇa, 46.11.
287. Ibid.
288. *द्वाभ्यां तु रथयन्तारं त्रिभिश्चास्य त्रिवेणुकम्।* Ibid., Droṇa, 131.47
289. Ibid., Ādi, 14.21.
290. Ibid.
291. Ibid., Sabhā, 2.14, 42.54; Virāṭa, 40.17.
292. Ibid., Bhīṣma, 23.14, 91.79.
293. Ibid., Udyoga, 23.14-15.
294. Ibid., Droṇa, 35.10; Bhīṣma, 50.64; Karṇa, 23.51.
295. Ibid., Virāṭa, 41.1.
296. Ibid., 13.13.
297. Śalya argued that *sūtas* were below Brāhmaṇas and Kṣatriyas in caste hierarchy. Ibid., Karṇa, 23.36.
298. *न त्वमर्हसिपार्थेन सूतपुत्र रणे वधम्।* Ibid., Ādi, 127.6.
299. E.W. Hopkins, op. cit., p. 253.
300. *MB*, Āraṇyaka, 19.3.
301. Ibid., Udyoga, 181.15.
302. *Rām*, Yuddha, 103.30-1.
303. *आदर्शयद् वासुदेवो हययाने परं बलम्।*
मोघान् कुर्वञ्शरांस्तस्य मण्डलानि विदर्शयन्।। *MB*, Bhīṣma, 102.48.
304. Ibid., Karṇa, 34.41.
305. *गतप्रत्यागतावृत्तैर्मण्डलैः सनिवर्तनैः।*
सारथेस्तु रथस्थस्य काश्यपेयस्य विस्मिताः।। Ibid., Droṇa, 122.52.
306. *देशकालौ च विज्ञेयौ लक्षणानीङ्गितानि च।*
दैन्यं हर्षश्च खेदश्च रथिनश्च बलाबलम्।।
स्थलनिम्नानि भूमेश्च समानि विषमाणि च।
युद्धकालश्च विज्ञेयः परस्यान्तरदर्शनम्।।
उपायानापयाने च स्थानं प्रत्यसर्पणम्।
सर्वमेतद् रथस्थेन ज्ञेयं रथकुटुम्बिना।। *Rām*, Yuddha, 104.18-20; also see *MB*, Karṇa, 28.6-8.
During the battle with Śālva, the charioteer of Pradyuma removed him from the battefield as he fell unconscious. *MB*, Āraṇyaka, 19.3-4.
307. *MB*., Karṇa, 67.
308. Ibid., Āraṇyaka, 23.21-2.
309. Ibid, Droṇa, 35.2-4.

310. *अधिष्ठितं हयज्ञेन सूतेनाप्तोपदेशिना। Rām*, Yuddha, 90.10.
311. *MB*., Udyoga, 183.5-6.
312. Ibid., Droṇa, 141.53-4.
313. Ibid., Śalya, 24.20.
314. Ibid., Bhīṣma, 50.105.
315. Ibid., Udyoga, 183.17-18.
316. Ibid., Bhīṣma, 71.30-1.
317. Ibid., Droṇa, 163.5-6.
318. *स्वयं सारथ्यमकरोत् पुनश्च धनुरस्पृशत्।*
तदद्भुतमभूत् तत्र सारथ्यं पश्यतां युधि।। Rām, Yuddha, 89.43.
319. *MB*, Āraṇyaka, 20.9; Sabhā, 50.16.
320. *वज्रसूच्यग्रं प्रतोदम्।* Ibid., Anuśāsana, 53.32; Ādi, 127.6.
321. *Rām*, Yuddha, 57.38.
322. *तस्य मेऽभ्याधिकं शस्त्रं सर्वोपकरणानि च।।*
संयुञ्जन्तु रणे क्षिप्रं शास्त्रवद्रथयोजकाः। MB, Śalya, 15.23.
323. *MB*, Karṇa, 22-3.
324. *अयं तु सदृशो वीरः शल्यः समितिशोभनः।*
सारथ्यं यदि मे कुर्याद् ध्रुवस्ते विजयो भवेत्।। Ibid., Karṇa, 22.50.
तथा शल्यसमो नास्ति हययाने हि कश्चन। Ibid., Karṇa, 22.56.
325. Ibid., Karṇa, 25.
326. *तानब्रवीत् पुनर्देवो मतः श्रेष्ठतरो हि यः।*
तं सारथिं कुरुध्वं मे स्वयं संचिन्त्यमाचिरम्।। Ibid., Karṇa, 24.96.
327. Ibid., Udyoga, 7.19-20.
328. *सारथ्यं तु त्वया कार्यमिति ये मानसं सदा।*
चिररात्रेप्सितं कामं तद्भवान् कर्तुमर्हति।। Ibid., Udyoga, 7.34.
329. *वामं चक्रं युधामन्युरुत्तमैजाश्च दक्षिणम्। MB*, Bhīṣma, 16.19.
330. *तं चक्ररक्षौ पाञ्चाल्यौ युधामन्युत्तमौजसो।* Ibid., Droṇa, 66.35; also 105.23.
331. *तस्यास्तां चक्ररक्षौ द्वौ सैन्धवस्य वृहत्तमौ।*
द्रौणिर्दक्षिणतो राजन् सूतपुत्रश्च वामतः।। Ibid., 70.49.
332. *द्रोणं यत्ताः पराक्रान्ताः सर्वे रक्षत पृष्ठतः।*
हार्दिक्यो दक्षिणं चक्रं शल्यश्चैवोत्तरं तथा।। Ibid., 139.18.
333. *शैनेयो दक्षिणं चक्रं धृष्टद्युम्नस्तथोत्तरम्।।*
पृष्ठगोपो भवत्वद्य मम पार्थो धनंजयः। Ibid., Śalya, 15.24.
334. *MB*, Udyoga, 152; The princes were also expected to be proficient in handling the chariot in the battlefield. In fact, it formed part of their training programme. *मार्गान रथचर्यासु। MB*, Ādi, 133.29 (BE).
335. *रथमार्गान् विचित्रांस्ते विचरन्तो महारथाः।*
प्रत्यदृश्यन्त संग्रामे शतशोऽथ सहस्रशः।। MB, Āraṇyaka, 170.34.
336. *चरन्त विविधान् मार्गान् रथेषु रथयूथपाः।* Ibid., Virāṭa, 31.18.

337. *ततो विराटस्य सुतः सव्यमावृत्य वाजिनः।*
यमकं मण्डलं कृत्वा तान् योधान् प्रत्यवारयत्॥ Ibid., Virāṭa, 52.27.
338. *मण्डलानि ततश्चक्रे गतप्रत्यागतानि च।*
यथा शक्ररथो राजन् युद्धे देवासुरे पुरा॥ Ibid, Droṇa, 18.6; also
मण्डलानां विभागेषु गतप्रत्यागतेषु च।
बभूव तुमुलं युद्धं तयोस्तत्र महामृधे॥
चरित्वा विविधान्मार्गान्मण्डलं स्थानमेव च।
शरैः पूर्णायतोत्सृष्टैरन्योन्यमभिजघ्नतुः॥ Ibid., Karṇa, 11.19-20.
339. *मण्डलानि विचित्राणि यमकानीतराणि च।*
चरन्तौ युद्धमार्गज्ञौ ततक्षतुरथेषुभिः॥ Ibid., 122.66 (BE).
रथाभ्यां चरेतुस्तत्र मण्डलानि सहस्रशः। Ibid., Karṇa, 18.71.
340. *अपसव्यं ततश्चक्रे द्रौणिस्तत्र वृकोदरम्।* Ibid., 11.17.
341. *प्रतिचक्रे च तं राजन् पाण्डवोऽप्यपसव्यतः।* Ibid., 11.18.
342. Ibid., 11.21-2.
343. *रथशिक्षां तु दाशार्हो दर्शयामास वीर्यवान्।*
उत्तमाधममध्यानि मण्डलानि विदर्शयन्॥ Ibid, Droṇa, 74.6
मण्डलानि विचित्राणि विचेरुस्ते मुहुर्मुहुः। Ibid., 74.15.; Also
मण्डलानि विचित्राणि गतप्रत्यागतानि च।
आदर्शयेतां बहुधा सूतसामर्थ्यलाघवात्॥ Ibid., Bhīṣma, 48.53.
मण्डलानि ततश्चक्रे गतप्रत्यागतानि च। Ibid., Droṇa, 18.6.
344. *अर्धमण्डलमावृत्य सूतपुत्रमयोधयत्।* Ibid., 106.20.
345. *गतप्रत्यागतावृतैर्मण्डलैः संनिवर्तनैः।* Ibid., 122.52.
346. *मण्डलानि विचित्राणि यमकानीतराणि च।*
सव्यानि च विचित्राणि दक्षिणानि च सर्वशः॥ Ibid., Āraṇyaka, 20.8;
अन्तरंच प्रहारेषु तर्कयन्तौ महारथौ।
राजन्नन्तरमार्गस्थौ स्थितावास्तां मुहुर्मुहुः॥ Ibid., Bhīṣma, 48.53-54.
347. *अपसव्यं चकाराथ माद्रीपुत्रस्तवात्मजम्।*
किरञ्शरशतैर्हृष्टस्तत्र नादो महानभूत्॥
अपसव्यं कृतं संख्ये भ्रातृव्येनात्यमर्षिणा।
सोऽमर्षितस्तमप्याजौ प्रतिचक्रेऽपसव्यतः॥
ततः प्रतिचिकीर्षन्तमपसव्यं तु ते सुतम्।
न्यवारयत तेजस्वी नकुलश्चित्रमार्गवित्॥ Ibid., Droṇa, 162.48-50; Also Vana, 30.13; Karṇa, 11.17.
348. *तौ मण्डलानि चित्राणि रथाभ्याम् सव्यदक्षिणम्।*
चरमाणावयुध्येतां रथशिक्षाविशारदौ॥ Ibid., Droṇa, 38.29.
349. *सोऽपसव्यां चमूं तस्य शाल्वस्य भरतर्षभ।*
चकार नातियत्नेन तदद्भुतमिवाभवत्॥
अमृष्णमाणोऽपसव्यं प्रद्युम्नेन स सौभराट्। Ibid., Āraṇyaka, 20.11-12.
यन्तारमस्य सहसा त्रिभिर्बाणैः समर्पयत्॥

चक्रेऽपसव्यं त्वरितः स्यन्दनेन जनार्दनः।। Ibid., *Droṇa*, 27.28; also see *Rām*, Yuddha, 106.15.

350. अर्धमण्डलमावृत्य सूतपुत्रमयोधयत्। Ibid., *Droṇa*, 106.20.
351. तं तथा कोष्ठकीकृत्य रथवंशेन मारिष। Ibid., 90.18.
352. अथैनं कोष्ठकीकृत्य सर्वतस्ते महारथाः। Ibid., 146.3; also see, Śalya, 22.83.
353. गन्धर्वमस्त्रमायच्छद्रथमायां च योजयत्। Ibid., Droṇa, 44.21.
    रथचर्यास्त्रमायाभिर्मोहयित्वा परंतप। Ibid., 44.24.
354. ततो दुःशासनः क्रुधो रथशक्तिं समाक्षिपत्। Ibid., 102.92.
355. *Rām*, Yuddha, 43.21.
356. Ibid., Virāṭa, 31.18.
357. Ibid., Śalya, 20.26; Āraṇyaka, 77.8; Droṇa, 106.21; *Rām*, Yuddha, 102.18.
358. समा निरुदकाकाशा रथभूमिः प्रशस्यते। *MB*, Śānti, 100.22.
359. Also see E.W. Hopkins, op. cit., p. 197; Chariot warriors used a weapon called *rathaśakti* to humble their opponents. *Rām*, Yuddha, 43.21.
360. शीघ्रगेन रथेनाशु स दूतः प्राप्य यादवान्। MB, Sabhā, 13.40.
361. इन्द्रसेनेन सहित इन्द्रप्रस्थमगात तदा।
    व्यतीत्य विविधान् देशास्त्वरावान् क्षिप्रवाहनः।। Ibid., 13.42.
362. Ibid., 2.14-16.
363. Ibid., 45.60-3.
364. गच्छ त्वं रथमास्थाय हयैर्वातसमैर्जवे।
    खण्डवप्रस्थमद्यैव समानय युधिष्ठिरम्।। Ibid., 50.58.
365. संह्रादनो राजरथो य इहास्मानुपावहत्।
    जैत्रो रथवरः पुण्यो मेघसागरनिःस्वनः।। Ibid., 61.5.
366. Ibid., 61.19-20.
367. Rām, Ayodhyā, 71.
368. Ibid., 83.5-6.

CHAPTER 5

Mahājanapada Period

The epics portray the importance of chariot in the life of the people in tune with the Vedic tradition. It seems that this continued in the Mahājanapada age, which lasted from about sixth century BC to fourth century BC, with some dilution. During this period, sixteen large states (Aṅga, Magadha, Kāśī, Kośala, Vajji, Malla, Cedi, Vaṁśa, Kuru, Pañcāla, Matsya Sūrasena, Assaka, Avantī, Kamboja and Gāndhāra) and ten autonomous clans (Śākyas, Bhagas, Bulis, Kālmas, Koliyas, Mallas (Pāvā) Mallas (Kuśinārā), Moriyas, Videhas and Licchavis) ruled over the country.[1]

Though by now, a new set of rulers had emerged, the military system of the country, based on the concept of *caturaṅgiṇī*, does not seem have undergone much change. Knowing that the chariot played a dominant role in the Mahābhārata and Rāmāyaṇa, and that the Mauryans maintained a powerful chariot corps it is improbable that the dynasties which flourished in between did not maintain a sizeable chariot force. Though the evidence on this is not as abounding as in the epics, the *Aṣṭādhyāyī*, Buddhist *Nikāyas* and the Greek sources furnish convincing evidence on the use of the chariot in the Mahājanapada period.

AṢṬĀDHYĀYĪ

The *Aṣṭādhyāyī* of Pāṇini, a composition of sixth century BC, refers to the wide use of chariots. It mentions three chariot types named after the draught animal yoked to it.[2]

Pāṇinī classifies the chariots in two categories, viz., *kad-ratha* and *parama-ratha*.[3] Agrawal interprets these types as inferior and superior respectively.[4] But, these should more appropriately be called the smaller type and the bigger type. Perhaps, the bigger type was used for carrying men and material while the smaller one was used by chariot warriors.

It is significant that Pāṇini refers to upholstering of the chariot,[5] a very common practice during the epic times. Leopard and tiger skins were used for the purpose and the chariots so upholstered were called *dvaipa* and *vaiyāghra* respectively.[6] The chariots upholstered with yellow blankets were called *pāṇḍu kambalī.*[7]

Patañjali, the composer of *Mahābhāṣya,* elaborates that these included horse-chariot (*aśvaratha*), camel-chariot (*auṣṭraratha*) and ass-chariot (*gardhavaratha*).[8] The size of the wheel depended on the draught animal yoked to the chariot.[9]

Though chariots constituted one of the four limbs of the army, these were occasionally integrated with other arms for operational purpose. In such cases, a combination of cavalry and chariots was called *rathikāśvārohan* and that of infantry and chariots as *rathikapādātam.*[10] The office of *ratha-gaṇaka* maintained a complete record of chariots.[11] The road built for the passage of chariots were called *rathapatha.*[12]

A significant reference to Indian chariots in early fifth century BC is met in *The Histories of Herodotus.* The Indian contingent which formed a part of Xerxes expeditionary force against Greece included chariots drawn by either horses or wild asses. This force of 300,000 men was defeated at Plataea in central Greece in 479 BC.

BUDDHIST SOURCES

In the Pāli literature, chariots are widely mentioned, particularly in relation to travelling, hunting and war. The *Jātakas,* which may be placed in the *Mahājanapada* time-frame, also refer to chariots. In the *Mahānāradakassapa Jātaka,* there is an interesting simile relating to a king and his chariot-maker, suggesting that chariot was very popular.[13] References to Indra's chariot are also met in some *Jātakas.*[14] *Vessantāra Jātaka* says that one king Sañjaya maintained an army of 60,000 men, including the chariots.[15] The story of a ceremonial chariot, sent out to identify a suitable king, is also repeated in some *Jātakas.*[16] Chariots were produced in large numbers and one prince Vesantara is reported to have gifted seven hundred chariots.[17] The *Mahājanaka Jātaka* refers to beautifully decorated *dīpa* and *veyaggha* chariots.[18]

In a Jātaka story, the chariot is symbolically described as follows:

> Thy body is called a chariot, swift and provided with the mind as a charioteer, having the abstinence from all injury as its axle, liberality as its covering, a careful walk with the feet as the circumference of the wheel, a careful handling with the hands as the sides of the carriage; watchfulness over the belly as the nave of the wheel, watchfulness over the tongue as the prevention of the wheel rattling. Its parts are all complete through truthful speech; it is well-fastened together by the absence of slander.[19]

In this connection, some other components of the chariot like ropes, carriage pole, umbrella, thong, yoke, goad and steeds are also mentioned.

In the Buddhist *Nikāyas*, the chariot appears as an important constituent of the *caturaṅgiṇī*, i.e. fourfold army.[20] A king was expected to be well-versed in chariot riding and fighting while mounted.[21] In fact, a Kuru king is reported as boasting his skill in handling chariots.[22] Rulers, no doubt led their armies in the battlefield mounting a chariot.[23] In the *Aṅguttaranikāya*, a ruler is stated to be preparing for an impending war and asking his *rathakāra* to make two strong wheels.[24] It is said that, a war chariot carried four persons and that chariots were arrayed separately on the wings when the troops were deployed in the battlefield.[25]

In the Buddhist literature, there are references to chariots made of gold, silver, iron, etc.[26] suggesting that wooden chariots were plated with various metals to look strong and beauiful. The road provided for the movement of chariots was called *rathiyā.*[27] *Ratthika* appears to have been an officer of a chariot unit, but *ratthika petanika* was perhaps a hereditary officer.[28] There is also a reference to a person called *rathapāla*, who figures as an important person.[29]

Expert horse-breakers were called *dakkha assadamaka. Assadamasārathī* was probably the coach, who trained the horses for being yoked to chariots. The best among them was known as *dakkha yogācariyo*, i.e. master of the craft.[30]

Many references to the use of chariot as a means of transport are made in Buddhist texts. This is evident from the following instances in the *Mahāvagga*:

1. A Licchavi *senāpati* went to meet Gautam Buddha mounting a chariot. He was accompanied by 500 chariots.[31]
2. A *mahāmātya* of Bimbisāra travelled to the house of one Meṇḍhaka accompanied by his fourfold army.[32]
3. A *gṛhapati* named Meṇḍhaka went to meet Buddha mounting a chariot, called Bhadrayāna.[33]

For a long and fast journey a relay of chariots was arranged. For the night journey torch-bearers accompanied the chariots.[34]

About the same time, when the Magadhan empire was in ascendance, its rulers maintained a strong chariot arm. During his battle, with the Licchavi king Ajātaśatru fixed large knives to the wheels of his chariots, to make them lethal in the battlefield. The device was called *rathamauśala.*

GREEK SOURCES

The information about the use of the chariot in the battlefield available in Greek sources, though it pertains to the fourth century BC, may hold good for *mahājanapada* age. These sources while discribing the invasion of India by Alexander refer to the chariot component of various armies encountered by him in India.

Alexander after defeating the Achaemenid ruler of Iran in the battle of Arbela in 331 BC, moved on to conquer north-western India. On the way, he defeated many tribal states, and finally crossed the Indus River near Ohind in the spring of 326 BC. Afer forging relationships with the rulers of Taxilā and Abhisāra, he marched against King Porus, who ruled over the territory across Jhelum, and had refused to submit. Alexander after a deceptive move crossed the river and confronted the armies of Porus, arrayed against him on the eastern side. The classical accounts refer to the use of chariots by Indian ruler in the battle of Jhelum:

1. After learning that Alexander has crossed the river, Porus sent his son, 'at the head of 2,000 men and 120 chariots' to stop his advance.[35] Aristobolus and Curtius, however, put the number of chariots in his contingent at 60 and 100, respectively.[36] In the skirmish that followed, the son of Porus was killed. As for the chariots, whatever their number 60 or 120, these were captured horses and all 'for they proved heavy in the retreat and useless in the action itself by having stuck fast in the clay'.[37]
2. After this debacle, Porus himself marched against Alexander, leading an army of 30,000 foot, 4,000 horses, 200 elephants and 1,000 chariots. Some sources put the number of chariots at 300. Poros deployed his chariots on the wings ahead of the infantry.[38]

In the army of Porus, chariots enjoyed a place of pride as attested to by the Greek sources.

> Its main strength lay in the chariots, each of which was drawn by four horses and carried six men, of whom two were shield-bearers, two archers posted on each side of the chariot, and the other two charioteers, as well as men at arms, for when the fighting was at close quarters, they dropped the reins and hurled dart after dart against the enemy.[39]

In the battle that followed, the army of Porus suffered heavily, as the chariots were broken into pieces and the drivers of the chariots were killed.[40] The reason for the debacle has been explained in detail:

> On this particular day, these chariots proved to be scarcely of any service for the storm of rain which, as already said, was of extraordinary violence, had made the ground slippery, and unfit for horses to ride over, while the chariots kept sticking in the muddy sloughs formed by the rain, and proved almost immovable from their great weight. Alexander on the other hand, charged with utmost vigour. . . . The fighting had now become hot everywhere, when drivers of the chariots rode at full-speed into the midst of the battle, thinking they could thus most effectively succour their friends. It would be hard to say which side suffered most from the charge, for the Macedonian foot-soldiers, who were exposed to the first shock of the onset were trampled down, while the charioteers were hurled from their seats, when the chariots,

in rushing into action jolted over broken and slippery ground. Some horses took fright and precipitated the carriages not only into the sloughs and pools of water, but even into the river itself. A few which were driven off the field by the darts of the enemy made their way to Porus.[41]

It is notable that the chariots used by Porus in this battle were large and heavy, drawn by four horses and mounted by six men. Even without any rain or storm or slippery ground, these chariots could not have faced the fast moving Macedonian cavalry.

3. On his way back, during the voyage down the River Indus, Alexander defeated Kṣudrakas (Oxydrakai) and Mālavas (Malloi). They made a joint front and assembled an army consisting of 90,000 foot-soldiers, 10,000 horses and 900 chariots to challenge Alexander. After suffering a defeat, one hundred ambassadors of these nations came to surrender before Alexander. They all rode in chariots and were men of uncommon stature of a very dignified bearing. Subsequently, these man presented Alexander with 1,030 chariots, each drawn by four horses, besides other things.[42]
4 Sambastai (Ambaṣṭhas), a democratic people, assembled an army of 60,000 foot-soldiers, 6,000 horses and 500 chariots to face Alexander. But they also suffered a defeat.[43]
5. Another reference to a powerful chariot corps is met in connection with Agrammes, the king of Prasians, or the people of eastern kingdom. His army comprised 200,000 infantry, 20,000 cavalry, 3,000 elephants and 2,000 four-horsed chariots.[44] Plutarch, however, puts the number of chariots in his army at 8,000.[45]

The evidence furnished by the *Aṣṭādhyāyī*, Buddhist *Nikāyas*, and the Greek sources conclusively proves that the chariot continued to play an important role in the life of the people, both in war and peace during the *Mahājanapada* age. Despite the debacle suffered in the battle of Jhelum it continued to maintain supremacy. The Mauryas maintained a separate department to look after the chariot corps.

NOTES

1. *Aṅguttara Nikāya*, 1.213; 4.252, 256, 260.
2. *Aṣṭādhyāyī*, 4.3.122.
3. Ibid., 6.3.102.
4. V.S. Agrawal, *India as Known to Pāṇini*, p. 149.
5. परिवृतो रथः। *Aṣṭādhyāyī*, 4.2.10.
6. द्वैपवैयाघ्रादञ् Ibid., 4.2.12; An early reference to *vaiyāghra-ratha* is met in the *Atharvaveda* 4.8.4.
7. Ibid., 4.2.11.
8. Ibid., 2.318.
9. Ibid.
10. V.S. Agrawala, *India as Known to Pāṇini*, p. 409.
11. *Aṣṭādhyāyī*, 5.1.129.
12. Ibid., 5.3.100
13. *The Jātakas*, vol. 6, p. 125.
14. *Kulāvaka Jātaka*, *The Jātakas*, vol. I, p. 80; *Sādhīna Jātaka*. Ibid., vol. 6, p. 224.
15. *The Jātakas*, vol.6, p. 298.
16. *Darimukha Jātaka*, Ibid., vol. 3, p. 157; *Sonaka Jātaka*, vol. 4, p. 128; *Mahājanaka Jātaka*, vol. 6, p. 25; also see vol. I, p. 80; vol. 6, p. 224.
17. *The Jātakas*, vol. 6, p. 503.
18. Ibid., pp. 48-50.
19. Ibid., vol. 6, p. 125.
20. *Dīgha Nikāya*, 3.200; *Majjhima Nikāya*, 3.173; *Saṁyutta Nikāya*, 1.73, 84; *Aṅguttara Nikāya*, 3.327; *Vinaya Piṭaka*, 1.241, 342; 2.10, 182; 4.105, 107, 108.
21. *Aṅguttara Nikāya*, 3.152.
22. *Majjhima Nikāya*, 3.177.
23. Ibid.
24. *Aṅguttara Nikāya*, p. 149.
25. *Vinaya Piṭaka*, 4.105-8.
26. Priya Srivastava, *Prācīna Baudha Granthon meṅ Varṇita Dhātu aur Dhātu-karma*, pp. 86, 87,149, etc.
27. *Vinaya Piṭaka*, 2.260
28. *Aṅguttara Nikāya*, 3.76, 78, 300.
29. *Vinaya Piṭaka*, 3.148.
30. Yashpal, *A Cultural Study of Early Pāli Tripiṭakas*, p. 183; Also see pp. 119, 131, etc.

31. *Mahāvagga*, 6.4.8.
32. Ibid., 6.6.2.
33. Ibid., 6.6.3; In the *Majjhim Nikāya*, there are many more references to chariot transport (pp. 1096, 1103, 1180).
34. Yashpal, *A Cultural Study of Early Pāli Tripiṭakas*, vol. II, p. 119.
35. J.W. M'crindle, *Ancient India, Its Invasion by Alexander the Great*, p. 101.
36. Ibid., 207, 308.
37. Ibid.
38. Ibid., pp. 102-103.
39. Ibid., p. 207.
40. Ibid., p. 107.
41. Ibid., p. 207; A similar fate was met by the Persian army during the encounter with Alexander near Nineveh. 'It was another huge medley of contingents, and it relied for its chief force upon that now antiquated weapon, the war chariot. Of these Darius had a force of two hundred, and each chariot had scythes attached to the wheels and to the pole and body of the chariot. There seems to have been four horses to each chariot and it will be obvious that if one of those horses was wounded by javelin or arrow, that chariot was incapacitated. Darius began the battle by flinging these instruments against the cavalry and light infantry. Few reached their objective and those that did were readily disposed of. H.G. Wells', *The Outline of History*, vol. I, p. 237.
42. Ibid., p. 234.
43. Ibid., p. 292.
44. Ibid., pp. 221-2.
45. Ibid., p. 310.

CHAPTER 6

Maurya-Śuṅga Period

There is a mistaken belief that after a disastrous performance in the battle of Jhelum, the chariot faded out of the Indian battlefield. On the contrary, there is substantial art and literary evidence to suggest that it continued to enjoy primacy in the Indian battlefield for many more centuries.

The Mauryan army comprised a strong chariot component. According to Greek sources based on the *Indica* of Megasthenes it included 6,00,000 infantry, 30,000 horses, 9,000 elephants and 8,000 chariots.[1] As each of the 8,000 chariots is presumed to have accomodated two fighting men and one driver, the Mauryan chariot corps may have included 24,000 men. Megasthenes, who served as a Greek ambassador in the court of Candragupta Maurya, says that among the six boards which managed the Mauryan army one exclusively looked after the chariot corps. There cannot be any doubt that chariots contributed substantially to the conquests of Candragupta.

Two early Tamil writers, Māmulanār and Paraṇar, who lived in second century AD, also refer to the Mauryan chariot corps. The former says that the Mauryan war chariots made their way as far south as Podiyil Hill in the Tinnevalley district:

> The newly-installed Mauryas, whose army contained a very large number of elephants, marched down to attack them; and their beautifully decorated cars wore down hill sides, making dark passages through which clear water flowed in torrents.[2]

In another context Māmulanār described Podiyil Hill as, 'worn by the bright rolling wheels of the cars of Mauryas, who marched towards the south'.[3] Parṇar Koranar, another Tamil poet refers to the cutting down of the Podiyil Hill to make a roadway for war chariots of the Mauryas.[4] But the evidence on chariots abounding in the *Mahābhāṣya*, the *Arthaśāstra*, the *Indica*, the *Smṛtis* and the art of Bharhut and Sāñcī is far more clinching.

MAHĀBHĀṢYA

The *Mahābhāṣya* of Patañjali dated in second century BC, is a dependable source of information on chariots during the Maurya-Śuṅga period. It includes chariots among the four limbs of the army[5] and says that these moved on an evenly laid broad road called *rathyā*.[6] It adds that a chariot warrior could march faster than a cavalry or infantry soldier.[7] The charioteer was called *savyeṣṭha*, as he occupied the seat on the left side of the chariot.[8] Patañjali used the terms *sūta, pranetā* and *prājita* for the charioteer.[9]

The chariots were upholstered (*parivṛta*) in various ways. Those covered with cloth were called *vāstra*, those with woollens, *kaṁbala* and those with skins, *cārmaṇa*. The chariots upholstered with yellow blanket called *pāṇḍukambalī* were highly valued and were known as *rājastaraṇa* or royal spread.[10] The chariots covered with leopard and tiger skin were called *dvaipa* and *vaiyāghra* respectively.[11]

Horses, camels and asses were also yoked to the chariots, perhaps depending on the need. These chariots were named *āśva, auṣṭra* and *gārdava* respectively.[12] Occasionally, quality bullocks called *rathya* were also yoked to the chariots.[13] The quality chariots were called *paramaratha* whereas ordinary types were called *kadratha*.[14] An assemblage of chariots was called *rathakaṭyā*.[15]

ARTHAŚĀSTRA

The wide use of chariots during the Maurya-Śuṅga age is best substantiated by the *Arthaśāstra*. Though composed about the beginning of the Christian era its references to chariots cannot, but apply to the Mauryan times.[16]

The *Arthaśāstra* says that each of the four arms of *caturaṅgiṇī*, i.e. infantry, cavalry, elephant corps and chariot corps should be placed under a superintendent (*adhyakṣa*).[17] The superintendent of the chariot corps (*rathādhyakṣa*) was reoponsible for establishing factories for manufacturing chariots.[18] These were produced in seven sizes, and designated as *deva-ratha, puṣya-ratha, saṅgrāmika-ratha, pāriyāṇika-ratha, parapurābhiyānika-ratha, abhiyānika-ratha* and *vainayikā-ratha.*[19] The largest chariot was ten *puruṣa* in height and twelve *puruṣa* in length.[20] As each successive type was one *puruṣa* less than the previous one in dimension, the smallest type was four *puruṣa* in height and six *puruṣa* in length. The chariots were made of wood.[21]

R.P. Kulkarni has given this dimensions of eight types of chariots in a chart as follows:

Chariot	Height		Width	
	Vitasti	Meter	*Vitasti*	Meter
Best chariot (*Uttama-ratha*)	10	2.288	12	2.736
Temple chariot (*Deva-ratha*)	9	2.052	11	2.508
Festival chariot (*Puṣya-ratha*)	8	1.824	10	2.288
War chariot (*Saṅgrāmika-ratha*)	7	1.596	9	2.052
Travelling chariot (*Pariyāṇika-ratha*)	6	1.368	8	1.824
Chariot for attacking enemy territory (*Parapurābhiyānika-ratha*)	5	1.34	7	1.596
Training chariot (*Vainayikā-ratha*)	4	0.912	6	1.368

Note: *Vitasti* is said to be a measure of length equivalent to 12 *aṅgulas* or 9 inches. *AS*, 2.20.10.

Of the seven types mentioned by Kauṭilya, the last five might have been used for military purpose, as these were smaller in size and lighter in weight. The superintendent was expected to be well-conversant with the management of weapons, armour and accoutrements and the appropriate use of charioteers, chariot-attendants and chariot horses in various assignments.[22] The Chariot was a prized commodity and an unexportable item. Any trader who dared to breach the law was penalized. The import of chariots was not, however, an offence.[23]

One of the important responsibilities of the chariot superintendent was to ensure proper care of horses. He was expected to follow the superintendent of horses (*aśvādhyakṣa*).[24] Only quality horses imported from the lands of Kāmboja, Saindhava, Āraṭṭa and Vanāyu were to be yoked to the chariots. The middling types came from Bāhlīka, Papeya, Sauvīra and Titala, while the others were considered inferior.[25] No doubt, some of these horse types were imported. The horses incapacitated by war, disease or age were looked after by the state properly.[26]

The best of the chariot horses could cover a distance of twelve *yojanas* in a day, the middling nine *yojanas* and the inferior six *yojanas*. Among the mounted horses, the best could cover ten *yojanas*, the middling seven and a half *yojanas* and inferior five *yojanas* in a day.[27] All the arms, including the chariots corps, performed necessary exercises at sun-rise outside the city. The king was expected to be present at these exercises to inspect their performance.[28]

The chief of the army excelled in mounting a chariot.[29] The king himself led the army mounting a chariot or an elephant.[30] In the capital city, roads were provided for the movement of chariots. These were made five *āratnis* (*hastas*) wide.[31]

As the chariots were widely used in the battlefield, tactics had been devised to contain their onslaught. The *Arthaśāstra* states that elephants, machines and carts, if well-equipped with stones, clubs, armour, hooks and hair-seizers and deployed at the centre these could prove a counter-force against the chariot divisions.[32] On the other hand, mailed chariots and armoured infantry were considered a powerful counterforce against the fourfold army.[33]

A suitable battlefield was necessary for the deployment of chariots. The *Arthaśāstra* describes the characteristics of such a battlefield as follows:

> Level, firm, clear, not causing jolting, not causing wheels or hooves to get stuck, not obstructing axles, not broken with trees, thickets, creepers, tree-stems, fields under water, pits, ant-hills, sand and mud, and free from clefts.[34]

At another place, it is said that the ground

> possessed of water reservoirs and shelters, not causing jolting, without fields under water, and enabling a turn round is excellent for chariot.[35]

The functions of chariots in the battlefield included the guarding of one's own troops, repelling the fourfold army in the battle, capturing, setting free, reuniting broken ranks, breaking up unbroken ranks, causing terror, showing magnificence, and making a frightful din.[36]

The modes or strategems, followed while using chariots in a battle, included rushing forth, rushing about, rushing beyond, rushing back, enclasping, moving zigzag, encircling, scattering, turning back after fleeing, guarding broken ranks along the lines, in front, on the flanks and centre, making a sudden assault, fighting in a stationary position on land suitable to them.[37] Mailed chariots were also deployed.[38]

In the battlefield, chariots were deployed at a distance of five *samas* (70 *aṅgulas*) from each other.[39] If considered necessary, this distance could be doubled or trebled to avoid crowding.[40]

A chariot unit was based on a single chariot and included five horses, fifteen soldiers and fifteen foot-guards. A chariot division, comprising three rows of three chariot units, at the centre with a flank and wing of the same size on either side, thus included 45 chariots, 225 horses, 675 soldiers and 675 foot-guards.[41] The division could be strengthened in size by two chariot units at a time, up to 21 units in all.[42] Still, if there were spare chariots, two-third of them could be deployed in the flanks and wings and one-third in the centre.[43] This was the general composition of a *sama-vyūha* described in the *Arthaśāstra.*

But in a *miśra-vyūha* (mixed array) chariots, elephants and horses, were deployed in the centre, flanks and wings, depending on the nature of the conflict. There were three types of mixed arrays. In the first type, called *viṣama pakṣabhedī vyūha* (array piercing through the wings), the chariots were deployed in the centre. In the second type, called *viṣama madhyabhedī vyūha* (array piercing through the centre), the chariots were deployed on the flanks. In the third type, called *viṣama antarbhedī vyūha* (array piecing through flanks), the chariot were deployed in the wings.[44]

An array with chariots at the centre, elephants on the flanks and horses in the rear was called *ariṣṭa* (unharmed). The array with foot-soldiers, horses, chariots and elephants deployed one behind the other was known as *acala* (immovable). The one in which elephants, horses, chariots and foot-soldiers were deployed one behind the other was called *apratihata* (unrepulsed).[45] The cart array when intermixed with elephants, horses and chariots was called *paripantaka* (flying about).[46]

The chariots were made of wood. There were forests (perhaps reserved), which provided material for manufacture of carriages and chariots.[47] *Rathopastha* (the body of the chariot) was considered sacrosanct. The rulers, while concluding a treaty sweared by touching the chariot, so that neither side dared its transgression.[48]

In view of a large chariot force, the kingdoms ensured that *sūtrādhyakṣa* (superintendent of yarns) produced bindings for vehicles and draught animals is sufficient quantity. These bindings were made of yarn and fibres and thongs of canes and bamboos.[49]

In the capital city, stables for chariots and other carriages were located in the north-west direction.[50] Separate roads, five *āratnis* in width, were reserved for chariots.[51] In a military camp, separate space was provided for lodging the chariots.[52] But the chariots meant for the personal use of the king were lodged just outside the audience hall.[53]

As for the emoluments of the officers of the chariot corps the commandant (*rathamukhya*) was entitled to a salary of 8,000 *paṇas*, the superintendent (*rathādhyakṣa*) 4,000 *paṇas* and a chariot fighter (*rathika*) 2,000 *paṇas*, per annum.[54] The king's charioteer was entitled to a salary of 1,000 *paṇas.*[55] The *Nītiprakāśikā*, a later work, also describes the pay structure of chariot warriors in detail.[56]

MEGASTHENES' *INDICA*

Megasthenes served as the Greek ambassador in the court of Candragputa Maurya. In his *Indica*, as quoted by later historians, he had recorded that a governing body consisting of six departments directed the military affairs of the state. One of these departments looked after the chariot corps. 'The chariots are drawn on the march by oxen, but the horses are led along by a halter so that their legs may not be galled and inflamed, nor their spirits dampened by drawing chariots. In addition to the charioteer, two men-at-arms sit besides him in the chariot.[57]

Taming of the horses to be yoked to chariots was indeed a demanding job. Such as undertake this work require to have a strong hand as well as thorough knowledge of horses. The greatest proficients test their skill by driving a chariot round and round in a ring; and in truth it would be no trifling feat to control with ease a team of four high mettled steeds when whirling round in a circle.[58]

In the Mauryan times, the care of the king's person was entrusted to women guards. Some of these guards moved on chariots.[59]

THE *SMṚTIS*

References to chariots in the *Smṛtis*, compiled about the beginning of the Christian era, also suggest the wide usage of chariots in the battlefield. *Manusmṛti* even advises that the rulers who were not sufficiently strong in chariot arm should avoid joining a war.[60] The chariot arm was considered so important that a ruler was expected to inspect his chariots along with other arms before retiring for the night.[61]

Chariots could not, however, be used in all kinds of terrain and during all the twelve months of the year. It was specified that chariots should be deployed appropriately only over an even terrain and only during the months of Hemanta and Śiśira (i.e. Agra-hāyaṇa to Phālguna).[62]

Interestingly, the chariots captured by a warrior during a campaign were considered his personal property and the state did not exercise any right over it.[63]

ARCHAEOLOGICAL EVIDENCE

Against the textual representations, archaeological evidence is almost non-existant about the presence of the chariots in early India. In fact, no remains of a chariot, even if they were in use, could have survived in Indian climate for a long time. The earliest chariot remains were discovered from Atrañjikheṛā dated between 350 and 50 BC. Most crucial evidence on the use of chariots during the Maurya-Śuṅga period comes from the Bharhut and Sāñcī *stūpas.*

BHĀRHUT

In Bhārhut *stūpa*, dated around 200 BC there are two representations of chariot. The first one, identified as the chariot of Prasenajit, is a two-wheeled vehicle with a high ornamental front and lower sides (Figure 6.1). It is of a large size, as it carries four persons. The king himself is shown as standing in mid-front with the driver on the left side, and an umbrella bearer behind. The chariot is drawn by four horses with plumes adorning their heads. Their long manes are plaited and their long tails are tied up on one side to prevent the animal whishing them on the charioteer's face.[64]

The second representation of the chariot is identical to the above-mentioned.[65] The dignitary mounting the chariot may also have been some king (Figure 6.2).

Vihāra No. 19 at Bhājā, represents Sūrya mounted in his chariot on the left side of the Varāṅḍā (Figure 6.3). The chariot appears to be trampling some kind of figure, probably the demon of darkness. There is also a view that it symbolises the trampling of the enemies of Buddhism.[66]

Figure 6.1: Chariot on Move – Bhārhut

Figure 6.2: Chariot – Bhārhut

Figure 6.3: Chariot – Bhājā

SĀÑCĪ

The wide representation of the chariot in Sāñcī reliefs, confirms its popularity during the Maurya-Śuṅga period.[67] In this connection, the relief captioned as the 'War of Relics' is of unusual interest (Figure 6.4). Marshall describes the scene is follows:

> to the left is a city, somewhat summarily represented, and sweeping towards its gateway a vast throng of people, soldiers and civilians, chiefs and their attendants, some on foot, some on horseback, others on elephants or in chariots, all surging onward to the accompaniment of drum and flute and conch-shell.[68]

In this scene the chariot is depicted as an integral part of *caturaṅgiṇī senā*. It also suggests that the kings and senior commanders went to war mounting chariots in the epic style.

It is notable that Sāñcī chariot is a moderately sized vehicle, with two wheels, pulled by two horses with sitting space for two persons, the warrior and the charioteer. The two-wheeled chariot is frequently represented on the gateway reliefs of Sāñcī (Figure 6.5). The number of spokes in the wheels is either 16, 20, 24 or 32. In the Sāñcī chariot, the tail of the horse is tied to a strap connected with neck-strap in the Bhārhut style so that it did not get entangled in the wheels or hit the charioteer in the face. It is notable that in Sāñcī chariot, horses are mostly controlled by reins fitted to the noseband, i.e. by pressure on the nostrils and occasionally by bit placed inside the mouth.

Figure 6.4: Battle of Relics – Sāñcī

Figure 6.5: Royal Procession – Sāñcī

It is likely that Sāñcī chariot was not much different from Vedic chariot. According to Sparreboom, 'there are no strong indications against the assumption that the Sāñcī chariots were basically very similar to those featured in Vedic texts'.[69]

In the sculptures at Amarāvatī[70] and Bodh-Gayā,[71] the soldiers mounted on chariots are depicted shooting arrows. But here, the construction of the chariot is very different from Bhārhut and Sāñcī chariot (Figure 6.6).

Figure 6.6: Battle Scene – Amarāvatī

The literary and art evidence corresponding to the Maurya-Śuṅga period of Indian history suggests that the chariot maintained its primacy in India even after its poor performance in the battle of Jhelum. Evidence furnished by the Greek sources and the *Arthaśāstra* as also the Bhārhut and Sāñcī art is conclusive in this regard.

THE ŚAKA-KUṢĀṆA PHASE

The centuries following the decline of the Mauryan empire witnessed large scale invasions and migrations leading to foreign occupation of northern India and parts of the Deccan. The predominance thus acquired by the foreigners such as the Yavanas, the Śakas, the Kuṣāṇas and the Pahlavas in the political life of the country lasted for a good four centuries. This political hegemony must have influenced the social and cultural life of the country in a fundamental manner.[72] But the influence might have been much more pronounced on military institutions as the invaders largely depended on their military prowess for survival.

The foreign invaders owed their victories in India to their cavalry and infantry. The Indians, who largely depended on their chariots and elephants, both for offence and defence, failed to measure up to the fast moving cavalry of the invaders, which comprised archers and javelin-bearers.

It seems that during this period of foreign domination over large parts of India, two sets of military formulations ruled over the Indian battlefield. While the native rulers followed the *caturaṅgiṇī* model, inherited from their predecessors, the invaders followed a new model, based on fast moving cavalry, particularly the mounted archer. As the latter model proved highly successful, the native model might have suffered some loss of prestige. In the process the chariot might have also suffered a jolt and indeed for some valid reasons:

i. The chariot could be deployed on a level ground and only in dry season.
ii. It was a comparatively slow moving vehicle and could not match the fast moving cavalry.

iii. In a battle, it was easier to manoeuvre a horse than a chariot.

But to say that the chariot eclipsed from the Indian battlefield as a consequence of foreign invasions and its poor showing would be far from truth. If that was really the case then the Indian elephant should also have suffered the same fate.

In this connection, it is notable that some Śaka and Kuṣāṇa rulers, after settling in India, also patronized the Indian *caturaṅgiṇī* model of military organization. Some silver and copper coins of Maues, an early Śaka king, represent the king riding a chariot pulled by two horses.[73] In the Junagadh inscription, Śaka ruler Rudradāman (AD 151) speaks highly of his *caturaṅgiṇī*, which included a chariot corps indeed.[74] He even boasts of his expertise in the management of horses, elephants and chariots.[75]

Menander, who ruled over some parts of north-west India in the second contury BC also maintained a fourfold army.[76] He went to meet Nāgasena, a Buddhist seer, mounted on a chariot. The seer convinced the Greek ruler about the composite nature of human personality through the simile of a chariot, which was composed of various parts like axle, wheels, frame, reins yokes, spokes, etc.[77] It follows that some other foreign rulers may also have maintained a chariot corps. In this connection, it is notable that representation of chariots in Mathurā art of Śaka-Kusāṇa period suggests that the chariot had not been discarded by foreign rulers.[78] In fact, a person mounting a two-wheeled chariot in a Mathurā sculpture has been indentified as Caṣṭana, a Śaka Kṣatrapa.[79]

In this connection the representation of Sūrya at Kaṇkāli *ṭīlā* in Mathurā is significant as the god is represented here in foreign costume. Two horses on either side of Sūrya, symbolically pull his chariot.[80]

It is interesting that the *Caraka Saṁhitā*, a medical treatise, compiled during this period also refers to many distresses (*rathakṣova*) caused by the jerks and shakes during a chariot ride. These included joint pains, pain in nose, ears and head, burning sensation in body parts, etc.[81] It also prescribed the treatment for these distresses.[82]

It is notable that a large part of Indian territory did not suffer the subjection of foreign invaders. The native rulers, might have generally followed their traditional model of *caturaṅgiṇī*, in which chariot and elephant played a dominant role.

NOTES

1. V.A. Smith, *Early History of India*, p. 132.
2. K. Aiyanger, *Epigraphia Indica*, vol. 8, p. 48.
3. Ibid.
4. Ibid.; There is a view that the Mauryas mentioned in ancient Tamil literature may have been the rulers of Koṅkan and not Magadha. But under no circumstances the petty rulers of Koṅkan could have maintained such a powerful chariot corps.
5. *हस्त्यश्वरथपदातिभिः। Mahābhāṣya*, 1.1.72.
6. *रथाय हिता रथ्या।* Ibid., 5.1.60.
7. Ibid., 1.1.70.
8. Ibid., 8.3.97.
9. Ibid.
10. Ibid., 4.2.10-11.
11. Ibid., 4.2.12.
12. *पतन्ति येनेति पत्रमश्वादिकं वाहनमुच्छते-अश्वरथं चक्रम औष्ट्ररथं गार्दभ रथम्।* Ibid., 4.3.122.
13. Ibid., 4.3.120.
14. Ibid., 6.3.102.
15. Ibid., 4.2.50.
16. According to Raychaudhuri the *Arthaśāstra* (*AS*) may have existed before second century AD. *Political History of Ancient India*, p. 226.
17. *AS*, 2.30-3; The concept of *caturaṅgiṇī* appears fully evolved in the *Arthaśāstra*. See *AS*, 1.19.15; 2.33.9; 5.3.35; 10.5.57.
18. *स रथकर्मान्तान् कारयेत्।* Ibid., 2.33.2.
19. *दशपुरुषो द्वादशान्तरो रथः। तस्मादेकान्तरावरा आ षडन्तरादिति सप्त रथाः।* Ibid., 2.33.3-4; According to *Śulbasūtra* of Baudhāyana (30.1) 1 *puruṣa* is made of 5 *āratnis* and each *ārtani* of 30 *aṅgulas* (3/4 inch); L.N. Rangarajan, however, considers 1 *puruṣa* equivalent to 9 inches. *The Arthaśāstra*, p. 699. Monier Williams considers 1 *ārtani* equivalent to 24 *aṅgulas*.
20. *AS*, 7.14.25.

21. Ibid., 2.33.5.
22. *इष्वस्त्रप्रहरणावरणोपकरणकल्पनाः सारथिरथिकरथ्यानं च कर्मरस्वायोगं विद्यात्।* Ibid., 2.33.6.
23. *AS*, 2.21.22-3.
24. Ibid., 2.33.1; also 2.30.
25. *प्रयोग्यानामुत्तमाः काम्बोजसैन्धवारट्टजवानायुजाः, मध्यमा वाह्लीकपापेयकसौवीरकैतलाः, शेषाः प्रत्यवराः।* Ibid., 2.30.29;
    As for the identification of these place names, Kāmboja was a kingdom to the north of Gāndhāra. Āraṭṭa was a part of Punjab. Vanāyu was perhaps Arabia or Persia and Vāhlīka was Bactria. Taitala was in Orissa and Sauvīra was adjacent of Sindhu. For details see R.P. Kangle, *The Kauṭilīya Arthaśāstra*, vol. II, p. 172 fn.
26. *AS*, 2.30.27.
27. *AS*, 2.30.39; Monier-Williams explains that *yojana* was the distance traversed in one harnessing or without unyoking; especially a particular measure of distance, sometimes regarded as equal to 4 or 5 English miles, but more correctly = 4 *krośas* or about 9 miles; according to another calcuation = 2½ English miles, and according to some = 8 *krośas*.
28. *AS*, 5.3.35-36.
29. *रथचर्या।* Ibid., 2.33.9.
30. *हस्ती रथो वा राजवाहनम्।* Ibid., 10.3.40; also 1.21.24.
31. *पञ्चारत्नयो रथपथः।* Ibid., 2.4.5.
32. *तदेव पाषाणलगुडावरणाङ्कुशकचग्रहणीप्रायं रथबलस्य प्रतिबलम्।* Ibid., 9.2.27.
33. *कवचिनो रथा आवरणिनः पत्तयश्च चतुरङ्गबलस्य प्रतिबलम्।* Ibid., 9.2.29.
34. *AS*, 10.4.3.
35. Ibid., 10.4.10.
36. *स्वबलरक्षा चतुरङ्गबलप्रतिषेधः संग्रामे ग्रहणं, मोक्षणं भिन्नसंधानमभिन्नभेदनं त्रासनमौदार्यं भीमघोषश्चेति रथकर्माणि।* Ibid., 10.4.15.
37. *AS*, 10.5.53-55.
38. *AS*, 9.3.29.
39. *पञ्चशमान्तरं रथम्।* *AS*, 10.5.3; *Śama* is a distance of 14 *aṅgulas*.
40. *द्विगुणान्तरं त्रिगुणान्तरं वा व्यूहेत। एवं यथासुखमसंबाधं युध्येत।* Ibid., 10.5.4-5.
41. Ibid., 10.5.9-13.
42. Ibid., 10.5.14.
43. Ibid., 10.5.20.
44. Ibid., 16.5.31-3; For a detailed study of the *vyūhas* in the *Arthaśāstra* see U.P. Thapliyal, *Warfare in Ancient India.*, pp. 230-51.
45. Ibid., 10.6.40-41.

46. Ibid., 10.5.29.
47. द्रव्यवन दुर्गकर्मणां यानरथयोश्च। *AS*, 7.14.25.
48. Ibid., 7.17.6.
49. सूत्रवल्कमयी रज्जूर्वरत्रा वैत्रवैणवीः।
    सांनाह्या बन्धनीयाश्च यानयुग्यस्य कारयेत्।। Ibid., 2.23.19.
50. पश्चिमोत्तरं भागं यानरथशालाः। *AS*, 2.4.12.
51. पञ्चारत्नयो रथपथः। Ibid., 2.4.5; *Āratni* was an elbow length; At another place the chariot path is said to be four *daṇḍas* (27 ft). चतुर्दण्डान्तरा रथ्याः। Ibid., 2.4.3.
52. Ibid., 10.1.7.
53. Ibid., 10.1.4.
54. Ibid., 5.3.9-12.
55. Ibid., 5.3.21.
56. *Nītiprakāśikā*, 7.33 42.
57. Strabo, 15.52, as quoted in *Ancient India as Described in Classical Literature* by J.W. M'crindle, p. 55.
58. J.W. M'crindle, *Ancient India as Described by Megasthenese and Arrian*, p. 89.
59. *Strabo*, 15.55, op. cit., p. 58.
60. *Manusmṛti*, 7.172; Also see 8.293-98.
61. Ibid., 7.222.
62. Ibid., 7.182, 192.
63. *Ibid.*, 7.96.
64. A. Cunningham, *The Stūpa of Bhārhut*, pl. 13.3.
65. Ibid., pl. 25.4.
66. Susan L. Huntington, *The Art of Ancient India*, Figs., 5.27, 28.
67. J. Marshall and A. Foucher, *The Monuments of Sāñcī*, vol. I, pls. 11.2; 15.3; 18a.2; 35a.1; 35b.2; 50 a.1; 51b; 60, 61.2,3; 62.2, etc.
68. Ibid., vol. I, pl. 61.2.
69. M. Sparreboom, *Chariots in the Veda*, p.10.
70. C. Sivaramamurti, *Amarāvati*, pl. 45 fig. 1.
71. B.M. Barua, *Buddha-Gayā*, fig. 42.
72. For details see U.P. Thapliyal *Foreign Elements in Ancient Indian Society*.
73. J.N. Banerjia, *Comp. History of India*, vol. II, p. 199.
74. *Epigraphia Indica*, vol. 8, p. 48; A chariot corps was always a component of *caturanginī*.
    हस्त्यश्वरथपदातं सेनांगं स्याच्चतुष्टयम्। *Amarakośa*, 2.8.33.
75. तुरग-गज-रथचर्या। Ibid.
76. *Sacred Books of the East*, vol. 25, p. 6.

77. Ibid.
78. R.S. Sharma, *Splendour of Mathurā Art and Museum*, 1950, pp. 56-7.
79. V.S. Agrawala, *Journal of the UP Historical Society*, 1949.
80. Susan L. Huntington, op. cit., p. 160.
81. *Caraka Saṁhitā, Sidhisthānam*, 12.12.2.
82. रथक्षोभातिचङ्क्रमणात्यासनजानां स्नेहस्वेदादि वातहरं कर्म सर्व निदानवर्जनं च। Ibid., 12.14.2.

CHAPTER 7

The Gupta and Post-Gupta Period

As discussed in the previous chapter, the archaeological evidence is scanty on the presence of the chariot during the Gupta age. However, in Ajāntā art, pertaining to this period, chariots are represented in the narration of some *Jātaka* stories. On the wall of the back aisle in cave XVII (fifth century AD), there is a painting on the left-hand side. The subject appears to be a battle scene as all four arms of the army are portrayed here. In the middle of the painting, a team of four horses pulling a chariot at a gallop is portrayed (Figure 7.1).[1]

The paucity of substantial art evidence has led scholars to believe that the chariot had ceased to play a role in Indian life by now. According to Chakrabarti, 'there is not an iota of evidence to prove that they (chariots) were employed by the Guptas, and it is certain that they were not used in the empire of Harṣa'.[2] However, there is abundant literary evidence to prove that the chariot

Figure 7.1: Chariot Ajāntā, Cave XVII

continued to play a part in the Indian life, both in war and in daily life, during the Gupta period. In this connection, the evidence furnished by Purāṇas, Nīti literature, *Amarakośa*, etc., is conclusive.

VIṢṆUDHARMOTTARA PURĀṆA

The *Viṣṇudharmottara Purāṇa*, dated around the fifth century AD, graphically describes the chariot, highlighting the following features:[3]

i. The chariot was made up of curved logs.
ii. It was provided with good wheels, tools and large banners.
iii. It carried quivers, bows and other weapons.
iv. It also carried *toya* (water), *jīraka (cumin)* and *piṇḍi* (food).
v. It was driven by an armoured charioteer, carrying a long whip.
vi. The charioteer was expert in holding the rein and in applying *maṇḍala* tactics.
vii. *Pārṣṇi*, the charioteer who managed the outer horses, knew the art of war and was an expert in wielding weapons.
viii. The charioteer carried a noose and a sword.
ix. Each chariot was occupied by two archers and charioteers each.
x. Chariot-archers were highly skilled and expert in wielding other weapons as well.
xi. The horses yoked to the chariot were well-accoutred and provided with light armour.

The chariots did play a significant role in the battles. In a battle-array, four chariots were deployed to protect an elephant and four horses to protect a chariot. Chariots were lined at the rear of the array ahead of the elephants.[4] An important function of the chariot corps was to cause a scare in the enemy ranks and the act was called *ratha-karma*.[5]

During a fierce battle between Bharata and Gandharva king Śailūsa, chariots were effectively used by both sides. In this encounter, many chariots were destroyed and many charioteers and chariot-horses were killed.[6] Again in a battle between *devas* and *rākṣasas*, the chariots and elephants destroyed each other.[7] The chariots pulled by elephants are also mentioned. The Goddess Dharā Devī is said to have joined the battle on the side of *devatās* mounting a chariot pulled by four elephants in the encounter between *devas* and *asuras*.[8] Further, a person who donated a chariot pulled by four elephants earned fruit equal to a donation of whole earth.[9] Further, it is said that the people of Sūraṣena and Pañcāla excelled in chariot war.[10]

MATSYA PURĀṆA

In the *Matsya Purāṇa*, a fifth-century AD composition, the gods and demons are described as fighting mounted on chariots. In fact, two chapters of the *Matsya Purāṇa* are devoted to describe the armies, including the chariot corps', of the contending adversaries.[11] Indra joined the battle mounting a chariot driven by Mātali.[12] The chariot of Sūrya was pulled by seven horses of green colour.[13] The chariot of the Moon god was a three-wheeled carriage pulled by ten horses of white colour.[14] The other planets also travelled in their chariots.[15]

The commanders of Tārakāsura's army are also described as mounting chariots pulled by lions, asses, horses, rhinos, elephants, etc.[16] Their chariots displayed standards on top.[17] The chariot of Maya comprised four wheels of iron and was drawn by asses.[18] It was decorated with golden nets, upholstered with rhino skin, carried a large stock of weapons and displayed *dhvaja* and *patākā*.[19]

In another context, it is said that both *devas* and *asuras* joined the battle mounting chariots and elephants.[20] There are graphic descriptions of chariot encounters in the *Matsya Purāṇa*.[21] Before joining the battle with the *daityas*, Śiva asked the gods to construct a special chariot for him so that he could lead them to victory. In this connection various components of the chariot including *kūbara, cakra, rathanemi, varūtha, triveṇu, yuga, aśva* (four), *pratoḍa, yoktra*, etc., are mentioned.[22]

There are also references to gifts of chariots with golden horses or elephants yoked to them. The procedure of gift is also detailed.[23] Purāṇic sources also describes the qualities of charioteer as knower of good and bad omens, expert in training horses, informed about the divisions of earth, good judge of merits and demerits of the warrior, a man of firm vision, sweet spoken and scholarly person.[24]

It seems that separate roads were provided for the movement of chariots. On special occasions, the chariot roads (*rathyāṣu*) were illuminated like the royal roads, palaces, houses,[25] and glittered like the heavenly roads in the moon-light.[26]

BHĀGAVATA PURĀṆA

In the *Bhāgavata Purāṇa*, there are many references to the use of chariot in the battlefield. When the Śālva king encircled Dvārakā and attacked it with lethal weapons Pradyumna led his band of *mahārathīs* to beat back the attack. In this encounter, when Pradyumna was seriously injured, his charioteer removed him from the battlefield to a safer place.[27] Pradyumna, on recovering, disapproved of this act of the charioteer as it amounted to cowardice on his part. The charioteer defended his act arguing that it was in accordance with the tenets of his *dharma* (duty) and that *rathin* and *sārathī* were mutually responsible to protect each other in the battlefield.[28] Pradyumna again joined the battle, which continued for 27 days. Later, Kṛṣṇa also joined the battle mounting his chariot.[29]

King Parīkṣita left his city for the conquest of four quarters mounting a chariot which was well-decorated, pulled by black horses and mounted with a lion standard. He was accompanied by an army of chariots, elephants, horses and foot-soldiers.[30]

During the battle between Kṛṣṇa and Jarāsandha at Mathurā, the army of Kṛṣṇa was completely encircled so much so that even the chariots became invisible.[31] Finally, Kṛṣṇa emerged victorious after destroying the army of Jarāsandha[32] and rendering his chariots useless.[33]

ŚIVA PURĀṆA

A graphic account of the divine *ratha* is met in the *Śiva Purāṇa*, wherein Sanatakumāra describes the constituents of Śiva's chariot in detail. Built by Viśvakarmā it was a two-wheeled vehicle of golden hue. Its right wheel with twelve spokes represented the sun, while the left wheel with sixteen spokes represented the moon. The other parts of the chariot included pole shafts, axles, pins of the yoke, flag staff, side trinkets, tinkling bells, pedestals, banners, steps, cords and chains, bridle, whip, umbrella, horses, charioteers, etc.[34] The great chariot was served by Brahmā as charioteer.

BRAHMĀṆḌA PURĀṆA

In the *Brahmāṇḍa Purāṇa*, there is a graphic description of the battle between Goddess Lalitā and Bhaṇḍāsura, the demon.[35] The Goddess joined the battle mounting the leading chariot called Cakrarāja, accompanied by many deities. While she occupied her seat called Ānanda-mahā-piṭha, the other goddesses, who were equal to her in grandeur, occupied various steps of the chariot. In this battle, she was also joined by Mantrīnātha seated on his Geyacakra chariot and Daṇḍanātha seated on his Kiricakra chariot. The other female deities occupied the steps of these chariots in the battle. In the midst of the armies of Śakti, these three chariots appeared like the three worlds in motion.

The three chariots, i.e. Cakrarāja, Geyacakra, and Kiricakra were respectively arrayed in the south-west, north-west and north-east direction of the Cintāmaṇi mansion to encounter Bhaṇḍāsura. In this connection, Cakrarāja, the Śrīdevī's chariot, is described in detail, though with some unnecessary exaggerations:

> It is very high with nine steps and is decorated with gems and jewels. The length and breadth are four *yojanas* each, and the height is ten *yojanas*. The pole to which the horses are yoked is thick and sparkling. The four Vedas are its great wheels. The four objectives of life are its great horses. It is adorned with *chowries*. It is rendered splendid by the pearl-studded umbrella. In the course of great battle with Bhaṇḍāsura it displayed adventurous exploits. The chariot was sanctified by Śrīdevī's seat.

In the *Brahmāṇḍa Purāṇa*, three is an important reference to the horses yoked to the chariot. These included Vanāyuja, Kāmboja, Pārada, Saindhava, Toṅkaṇa, Parvatīya, Pārasīka, Ājaneya, Darada, Yavana, Gāndharva and Kairāta.[36] Many of these horse breeds, which were undoubtedly imported, find mention in the epics as well.[37]

VĀYU PURĀṆA

In the *Vāyu Purāṇa*, there are many references to chariots. Rudra is said to have given a chariot to King Yayāti, which was radiant, unstoppable, divine, indestructible and was pulled by steeds, which ran faster than wind. Mounting this chariot, he conquered the earth and remained invincible to gods, demons and men. This chariot remained a prized possession of Paurava kings for generations.[38] The king Āhuka is said to have maintained an army of ten thousand chariots, all equipped with quivers, weapons, banners, etc.[39]

Again in connection with the brave deeds of Kārtavīrya Arjuna, the chariot arm is mentioned.[40] His five sons are stated to have been great chariot warriors.[41] King Yayāti also used a chariot during his campaigns.[42]

The *Vayu Purāṇa* also describes different types of roads on which the chariots plied. The *rājapatha*, which was 10 *dhanuṣa* (40 *hastas*) wide, was considered ideal for the movement of chariots, elephants, horses, etc. The other roads (*rathyā*), on which the chariots plied included *sākhārathyā*, *gṛharathyā*, *uparathyā*, and *uparathyakā*. Of these *sākhārathyā* was four *dhanuṣa* wide, while the others were two *dhanuṣa* wide.[43]

MĀRKAṆḌEYA PURĀṆA

During an encounter with the Goddess (Devī) the leading commanders of Mahiṣāsura are stated to have joined the battle with large chariot contingents as follows:[44]

Cāmara	fourfold army including chariots
Udagra	60,000 chariots
Mahāhanu	1,00,00,000 chariots
Asuloma	5,00,00,000 chariots
Bāṣkala	60,000 chariots
Ugradarśana	1,00,00,000 chariots
Viḍāla	5,00,000 chariots
Kāla	5,00,000 chariots

The armies of other Asura commanders also included considerable chariot component.[45] Though the number of chariots is totally imaginary, these certainly prove the importance of chariot in the Indian battlefield.

Many of these chariots were destroyed by the Goddess and lay strewn over the battlefield.[46] There are some more references to chariot encounters in the *Mārkaṇḍeya Purāṇa.* Dama, the son of Nariṣyanta, during his encounter with Vapuṣmān, the king of Kuṇḍina, cut of the head of the latter's charioteer[47] and forced him to flee.[48] After he had won the hand of Sumanā, the daughter of the ruler of Dasārṇa, his father-in-law presented him with many chariots.[49] Later, in another encounter with Vapuṣmān he asked his minister to organize his fourfold army appropriately.[50] In the ensuring battle, the chariots fought with the chariots only.[51] In this contest princes were expected to display the skill of fighting mounted on chariots.[52]

AGNI PURĀṆA

In the *Agni Purāṇa*, the chariots of various gods are described in detail. The chariot of the Sun is said to be one-wheeled with five spokes (*ara*) and six felloes (*nemi*). It was pulled by seven horses.[53] The chariot of the Moon, which was three-wheeled, was pulled by ten white horses.[54] The chariots of planets like Buddha, Bṛhaspati, Śani, Rāhu and Ketu were pulled by eight horses each.[55]

There are many references to chariots in relation to war. It is said that, if a war is waged between mid-November (Hemanta) and mid-March (Śiśira) the army should include a strong chariot corps and cavalry. However, in the event the war was waged during the months of Vasanta (spring) and Śarada (autumn) all four arms should be evenly matched.[56]

A level battlefield was considered good for the deployment of chariot corps.[57] It is further explained that the terrain which is not covered by stumps, trees, hills and mud is best suited for the chariots to be engaged.[58] In the scheme of *vyūhas* described in the *Agni Purāṇa*, the chariots were appropriately deployed in the flanks and wings.[59] In the order of battle, they were placed behind swordsmen, bowmen, horsemen and ahead of the elephants.[60] In the battlefield four horsemen served as guards to a chariot.[61] At another place it is said that, three archers were deployed for the protections of one chariot.[62]

In a battle the main task of the chariot corps was to disperse the enemy troops and rally its own troops when in disarray.[63] Chariots were also employed for removing the critically injured from the battlefield and for harassing the enemy troops.[64]

SKANDA PURĀṆA

In the *Skanda Purāṇa*, there are many references to the chariots in the context of wars between the *devas* and *dānavas*. When Mahiṣāsura's love was contemptuously rejected by the Devī, he felt very offended and asked his *senāpati* to prepare his strong four-fold army comprising elephants, horses, chariots and infantry.[65] This army was to include one lakh elephants, three lakh chariots, ten lakh horses and innumerable foot soldiers. The chariots were decorated with gems.[66]

DEVĪBHĀGAVATA PURĀṆA

In the *Devībhāgavata*, there are references to the use of chariots in battles. During a battle between Yudhājita and Vīrasena, chariots were used alongside the other arms.[67]

But chariots were more frequently used for transport, one queen Manoramā left for her father's place mounting a chariot, accompanined by a minister and a lady servant. But on the way, Naṣādas looted her wealth and chariot.[68] After Śaśikalā, daughter of King Subāhu, was married to Sudarśan, the king presented him with two hundred chariots.[69]

ĀDI PURĀṆA

According to the *Ādi Purāṇa*, a Jaina composition, presumably compiled in eighth century AD, chariots were widely used in war. King

Bharata, when he went out on a campaign, mounted a chariot comprising two wheels and a strong axle. It was pulled by two horses and driven by a charioteer.[70] In the marching order, foot soldier moved first followed by horses, chariots and elephants.[71]

The groups of chariots, well-stocked with weapons and armour, looked like armouries[72] and were deployed on both flanks of the array.[73] In an encounter between Jayakumāra, son of Bharata, and his adversaries during the *svayaṁvara* of Sulocanā the chariots were used in large numbers by both sides.[74] Many other references to chariots in the *Ādi Purāṇa* suggest that these were widely used both in warfare and daily life.

In the *Pariśiṣṭaparvan* of Hemacandra, there are many instances of chariots being used for transport purpose:

1. While passing through a forest Valkalacārin met a *rathin*, who was accompanied by his wife (1.173).
2. Ṛṣabha, a big merchant, journeyed to Vaibhāra hill for recreation and rest along with his wife Dhāriṇī mounting a chariot (2.20-1).
3. On his way to meet Svāmī Sudharmā, prince Jambu travelled on chariot up to the city gate and walked on foot thereafter.[75]

The occurrence of the terms like *ratha-nūpura, rāthino-rātha, ratha-varya, ratha-vega, rathojitam, rathoma, rathino, rathaka*, etc. comfirm the wide use of chariot around eighth century.[76]

It is significant that Indians continued to import good quality horses from the adjoining countries even then. This is amply evident from Purāṇic literature.[77]

CHARIOTS OF PLANETS IN PURĀṆAS

SUN (*SŪRYA*)

In the *Vāyu Purāṇa*, the movements of celestial bodies like sun, moon and planets are described.[78] In this connection the chariot of planet Sun is described in some detail. It comprised one wheel with five spokes (*arāh*) and three naves.[79] The other components of his chariot included body (*nīḍa*), shafts (*ayana*), ropes (*bandhura*), yoke-kins (*kalāh*), shaft (*kaṣṭha*), axle tree or bottom (*nimeṣa*), frame (*īṣā*), fence (*varūtha*), standard (*dhvaja*), linchpins (*yugākṣakoṭi*), pole (*akṣa*) and bridle (*raśmi*).[80] The chariot was pulled by seven horses[81] of green colour.[82]

It may be mentioned here, that of all the divinities Sūrya or the Sun deity is most intimately associated with the chariot in Indian mythology. In the *Viśvakarmā-Śilpa*, the chariot of Sun is called Makaradhvaja, pulled on one wheel by seven horses.[83] *Liṅga Purāṇa*, describing the sun chariot in detail says that it was created by Brahmā as a single-wheeled vehicle with five spokes and three naves and six rims.[84]

There is an interesting reference to *rathayātrā* in the *Sāmba Purāṇa.*[85] Sāmba, the son of Jāmbavatī by Kṛṣṇa, was cursed to turn into a leper by his father. It is said that he got rid of the affliction by organizing a *rathayātrā*. The *Bhaviṣya Purāṇa* refers to a *rathayātrā* festival organized in Ujjayinī in the honour of Sun god. Participation in this festival was considered very rewarding and yielded the fruit of an Agniṣṭoma sacrifice.[86]

The Sun temple at Koṇārka, built around 1240 AD has been designed as a massive chariot. At Purī, the Jagannātha *ratha* constructed during the great *rathayātrā* festival follows the design of Sun Temple.

Moon (Śaśi)

The chariot of the moon was a three-wheeled vehicle with a hundred spokes each. It was pulled by ten white horses and is said to have sprung from the water. It was driven by Cakṣuśṛvarā, the charioteer.[87]

Mercury (Budha)

It was a shining vehicle yoked with eight tawny horses. The chariot was provided with a *varutha* (fence) and *anukarṣa* (axle) and was equipped with a quiver, banner and standard.[88]

Venus (Bhārgava or Śukra)

It was again a shining vehicle pulled by ten horses of different colours. The horses were sturdy and could run at the speed of the wind.[89]

Mars (Maṅgala or Kumāra)

The chariot of Kumāra was glorious and golden. It was yoked to eight red horses of exceptional speed. It is said that these horses emerged from the fire. On top the chariot bore a *dhvaja*.[90]

Jupiter (Bṛhaspati)

The chariot of Bṛhaspati was golden and was pulled by eight horses of red colour. The horses could compete with the speed of the wind.[91] In the *Matsya Purāṇa*, the chariot is described as made of silver and yoked to horses white in colour.[92]

Saturn (Śanaicara)

The chariot was made of black iron and yoked to white horses. It sprung out of the environment and moved at a slow speed.[93]

Rāhu (Svarabhānu)

The chariot was pulled by eight dark coloured horses, who could run at the speed of the wind.[94]

Ketu (Dānava)

The chariot was pulled by eight horses, smoky in colour. Strong and tawny like donkeys they could run at the speed of the wind.[95]

Before ending with the Purāṇas, it may be stated that there is a general tendency to discard the Purāṇic evidence, branding it as mythical, even without reading them. In fact, a lot of historical material is enshrined in the Purāṇas, which if critically examined may lead to astounding results, as in the case of archaeological finds. As for the chariots, nobody can deny the fact that these existed and were very much in use during the times, when Purāṇas were compiled.

Rajendra Lal Mitra seems to have erred in interpreting the evidence contained in the Purāṇas, while stating that 'their descriptions lead me to suppose that their idea of vehicle was the *ratha* of the modern times. A tower like structure of many storeys, a cumbrous apparatus altogether unfit for warlike purpose'.[96] In fact, there are few references to such big chariots in the Purāṇas in comparison to normal chariots, which could be used for travelling and in war. The literature compiled during the Purāṇic times also confirms this.

WORKS OF KĀLIDĀSA

Chariots widely figure in the works of Kālidāsa like *Raghuvaṁsa*, compliled during Gupta times. King Dilīpa went to the hermitage of Sage Vasiṣṭha, mounted on a chariot.[97] King Raghu slept in his chariot along with his weapons on the night preceding the march to the battlefield.[98] Prince Aja journeyed to attend the *svayaṁvara* of Indumatī, the daughter of Vidarbha ruler, in a chariot, at the head of a large army including chariots.[99] King Duṣyanta went on a hunting spree mounting a chariot driven by a charioteer (*sūta*).[100] King Pururavā is described as mounting a chariot driven by a *sūta.*[101] The chariot of King Vikrama was called Somadatta, and it carried a *dhvaja* bearing the symbol of *hariṇa* (*deer*).[102] The chariot of Indra was drawn by green coloured horses and driven by a *sārathi.*[103]

In the works of Kālidāsa, the chariot also figures as an important component of the army and an effective instrument of war. The army of Raghu included a chariot corps.[104] The army of prince Aja also included a chariot corps which participated in the battle he waged against the kings, who wanted to kidnap princess Indumatī.[105] Indra invited Duṣyanta to assist him in his war against the Dānavas and sent his chariot to fetch him. He indeed joined the battle mounting the chariot of Indra and defeated the foes.[106] Daśaratha led his army to victory mounted on a chariot.[107]

A charioteer, called *sūta* or *sārathi*, was invariably employed to pilot the chariot. But in the event he was killed in the battle the warrior (*rathin*) himself picked up the reins and continued the fight.[108]

The dramatist Bhāsa, who was a contemporary of Kālidāsa, also refers to chariots in his dramas:

1. Vatsarāja travelled to the court of Mahāsena in a chariot pulled by asses (*khara-rathen*).[109]
2. Chariots and charioteers are repeatedly mentioned in relation to transport.[110]

3. That chariots participated in the battle is evident from a battle account.[111]
4. The horses of Kāmboja were famous and were compared with the horses of the Sun God.[112]

KĀMANDAKĪYA NĪTISĀRA

In the *Kāmandakīya Nītisāra* the chariot corps is stated to be a vital component of sixfold army (*ṣaḍāṅgabala*).[113] Its importance is further underlined by the fact that the king himself supervised the repairing of the chariots.[114] In a battle formation the chariots were deployed at a distance of five *hastas* from each other and in such a manner that there was no obstruction in the event of a retreat.[115]

In ancient India, chariots were deployed in a *vyūha* (array) before joining a battle in the centre, on the flanks or on the wings depending upon the type of the *vyūha* decided upon by the commander. To illustrate the point, the *vyūha* in which infantry, cavalry, chariots and elephants were deployed one behind the other was called *acala-vyūha*. The array is which the cavalry was placed in the centre, the chariots on the flanks (*kakṣas*) and elephants on the wings (*pakṣas*) was called *antavid.*[116] While negotiating a danger prone area during a march the chariots were deployed on the flanks of the horses, while elephants were deployed on the flanks of the chariots.[117]

Kāmandaka Nīti considers the terrain, 'devoid of sandy soils, mud, earthen mound, gravels and stones and free from marshes, creepers, pits, trees, crops and such like things . . . ideal chariot grounds.'[118] It further adds that the ground for the deployment of chariots, horses and elephants should be steady and hard.[119]

ŚUKRANĪTI

The *Śukranīti*, considered to be a later work by some scholars, contains many references to chariots. In the capital city, space was provided in the north for keeping chariots, horses, weapons, etc.[120] A ruler was expected to practice the mounting of a chariot every morning.[121]

A charioteer was supposed to know all about the horses, their capacity to carry loads, the strength of the chariot and its various moves. Rendering the enemy attack ineffective by rapid moves and ensuring the safety of his horses during encounters was also considered a virtue of the charioteer.[122]

Śukranīti considered chariot making as an art[123] and therefore, advises the king to employ chariot-makers.[124] The deployment of various arms in war was again considered an art[125] and it was advisable to encounter a chariot with a chariot only.[126] It was also advisable for a king to build well protected beautiful sheds to the north of the capital to house chariots and horses.[127]

Mounting a chariot was a privilege enjoyed by a few. The commander of 1,000 troops enjoying a fief of 100 villages could travel in a chariot drawn by one horse,[128] the commander of 10,000 troops enjoying a fief of 1,000 villages could travel in a chariot drawn by two horses[129] and a commander of 50,000 troops enjoying a fief of 100,000 villages could travel in a chariot drawn by four horses.[130]

AMARAKOŚA

The *Amarakośa* has described chariots in some detail. It refers to seven types of chariots, viz., *śatāṅga, syandana, ratha, puṣyaratha, karṇīratha, pravahaṇa* and *dayana.* Of these, the first three types were indeed war chariots.[131] The fourth one, viz., *puṣyaratha* was not used in war[132] and its use was perhaps restricted to ceremonials. The last three chariot types were meant for the use of women.[133]

The chariots were upholstered to strengthen and adorn them. Those upholstered with tiger skin were called *dvaipa* or *vaiyāghra*[134] and these might have been used in war. *Syandana* again a war chariot, was covered with a yellow blanket and hence called *pāṇḍukaṁbalin.*[135] The chariots covered with blankets and clothes were called *kāmbala* and *vāstra* respectively.[136]

Amarakośa also refers to various parts of the chariot such as *dhura* or *yānamukha, cakra or rathāṅga, nemi* or *pradhi, nābhi* or *piṇḍikā, aṇi, varūtha* or *rathagupti, kūbara* or *yugandhara, anukarṣa, yuga* or *prāsaṅga,* etc.[137] These components are also mentioned in the epic.

The carpenters, engaged in making the chariots called *rathakāra*,[138] did not command much respect in the society.[139] It seems that chariots were made of the wood of *tiṁśa* tree, which was also called *syandana, nemi, rathadru*, etc.[140] The charioteer was called variously as *niyantā, prājitā, yantā, sūta, kṣattā, sārathi, saveyṣṭa* and *dakṣiṇastha*.[141] The last two names were ascribed to the charioteer because he sat on the right side of the *rathin*. The warrior who fought mounted on a chariot was called *rathin* and *syandanāroha*.[142] The horse yoked to the chariot was called *rathya*.[143]

Hemacandra, in his Sanskrit vocabulary, enumerates eight kinds of chariots, viz., *śatāṅga-ratha, syandana-ratha, puṣpa-ratha, maruda-ratha, yogya-ratha, parighātika-ratha, karṇi-ratha*, and *rathagarbhaka*. Of these chariots, the first two were intended for use in the battlefield, the third for enjoyment, the fourth for carrying the images of gods, the fifth for the use of magistrates, the sixth for travelling, the seventh for fighting and the eighth for ascending in the air.[144] The make and shape of these chariots is not described but some of them also find mention in the *Amarakośa*.

MĀNASĀRA

In the *Mānasāra*, a famous treatise on Indian architecture, dated between 500-700 AD a chapter in devoted to the chariots. After detailing the characteristic features and measurements of chariots suitable for the ride of the gods, the Brāhmaṇas and the kings it says:[145]

1. Of the aforesaid measures, the smaller varieties should be employed in the chariots of gods, the Brāhmaṇas and the kings for fighting and mock fighting, the intermediate varieties for daily use and the large varieties for special occasions (43.129-31).
2. All the chariots suitable for fighting should be furnished with three wheels; all those meant for mock-fighting or ordinary ride should be furnished with four wheels; those meant for daily use should be furnished with five wheels; and all the chariots meant for special use should possess six to ten wheels, depending on the size (43.132-6).

3. The chariots for fighting should have three platforms, those for mock-fighting two platforms and those for daily use one to three platforms. The chariots for special use may have three to nine platforms (43.137-9).

It follows from the above that the ceremonial chariots were made large sized and were used for carrying the images of gods and the kings. The use of chariots was common for travelling, particularly by Brāhmaṇas. In wars small chariots with three wheels and three platforms were used.

Besides, the above there are also some incidental references to chariots in the *Mānasāra* suggesting their wide use:

1. Chariot was a fast conveyance (*Man*, 3.9).
2. After building a house the owner entered it with pomp accompanied by elephants, chariots, etc. (*Man*, 37.86-9).
3. In a village the houses of charioteers (*sūta-kāra*) should be built in the south and south-west (*Man*, 36.43).
4. The warriors who fought mounting chariots, elephants and horses lived in *caturmukha* type of mansion (*Man*, 35.32-5).
5. While planning a village large streets (*rathyā*) for the movement of chariots were provisioned (*Man*, 9.192).

Rathamārga or *rathyā* is also mentioned in the *Aparājitapṛchhā*, a composition of twelfth century, suggesting that chariots were still in use.[146]

ŚIŚUPĀLAVADHA

The *Śiśupālavadha*, a work compiled in eighth century, refers to wide use of chariots for transport and in war:

1. Fast running chariots were brought to the level ground from the sliding space of the gate by the charioteers without much effort by pulling the reins of the horses tied to the yoke.[147]
2. A large number of chariots accompanied Kṛṣṇa during his march to Indraprastha.[148] These chariots cut through the stony path with their wheels,[149] suggesting that the outer rim of the chariot-wheel was made of metal, like iron.

3. The fast moving chariots raised considerable dust[150] and made a loud noise.[151]
4. Carpenters (*takṣaṇā*) accompanied the army for on the spot repair of the chariot.[152]
5. In a battle chariots engaged only chariots.[153]

There are many references to the chariot of Kṛṣṇa in the *Śiśupālavadha*. It was named Puṣya and could move in all directions at a rapid pace.[154] On top of the *dhvaja* staff sat Garuḍa.[155] The chariot was driven by Dāruka.[156]

NĪTIPRAKĀŚIKĀ

The evidence furnished by *Nītiprakāśikā* is far more substantive in this regard. This work dated around tenth century AD, not only mentions various grades of chariot warriors but also the salary payable to each of them in cash:[157]

Yuvarāja	5,000 *varvas* (a kind of coin)
Sarvasenapraṇetra	4,000 "
Atiratha	3,000 "
Mahāratha	2,000 "
Rathika	1,000 "
Ardharatha	500 "
Ekaratha	300 *niṣkas* (a kind of currency)
Sārathi	15 *varvāṇa*[158]

Nītiprakāśikā also specifies the responsibilities of the chariot corps in the battlefield. These included the protection of elephants, shooting the enemy from a distance to dislodge him, confronting the enemy while positioned at the entry point (*vyūha-dvāra*) and making loud noise.[159]

SAMARĀṄGAṆA-SŪTRADHĀRA

The most authentic evidence on the wide use of chariot in India is furnished by the *Samarāṅgaṇa-Sūtradhāra*. It is acknowledged to be a standard treatise on the Indian architecture and is ascribed to King Bhoja, who ruled over Malwa in eleventh century AD. Those inclined to dismiss the evidence furnished by the Purāṇas as mythical may find it difficult to challenge the historicity of this work.

This great treatise on architecture spread over 83 chapters (*adhyāyas*), refers to chariots, chariot roads, chariot construction, chariot use, chariot types, chariot deployment, etc. The chapters dealing with the planning of townships (*pura-niveśa),* royal palaces (*rājaniveśa*) and buildings (*prāsādas*) and manufacture of instruments (*yantravidhāna*) are particularly replete with references to chariots. Some of these references are illustrated below:

1. In one of the chpaters in *puraniveśa*, there are many references to chariot roads, big and small. Chariot roads (*rathyā*) starting from the palace area went up to the borders of the state.[160] This was done to facilitate the movement of army (*balasya caturaṅgasya*) to meet any threat on the borders.[161] The laying of chariot roads and their dimensions are also described in detail.[162]
   Further, in planning the township space was to be provided to the experts in handling chariots and wielding weapons, on the western side.[163]
2. In the *rājaniveśa*, it is said that shed should be built for the lodgement of chariots and elephants to the south of the royal palace.[164]
3. While setting up a royal camp (*śivira*) special care was taken in selecting the site and providing for chariot-road for the movement of the army.[165]
4. In *yantravidhāna*, the construction of various types of chariots is described in detail.[166]
5. In chapters describing the construction of various kinds of buildings (*prāsādas*), the chariot factor is always there as a decorative motif.[167]

ABHIDHĀNA CINTĀMAṆI

Hemacandra in the *Abhidhāna Cintāmaṇi* has classified chariots in six different categories, improving upon the classification met in the *Arthaśāstra* and *Amarakośa.*[168] These included: war chariot (*śatāṅga, syandana, ratha*), sports chariot (*puṣparatha*), divine chariot (*marudaratha*), training chariot (*yogyāratha vainayika*), travel chariot (*aśvaratha, pariyāṇika*), chariot for women (*karṇī-ratha, pravahaṇa, dayana, rathagarbhaka*).

For protection, chariots were wrapped with blanket (*kambala*), cotton cloth (*vāstra*) and leopard (*dvaipa*) and tiger (*vaiyāghra*) skin.[169] In this connection various parts of chariot have also been mentioned.[170]

MĀNASOLLĀSA

Some irrefutable evidence on the existence of chariot in twelfth century AD is furnished by the *Mānasollāsa* compiled by Someśvara. It says that a chariot should be fully equipped with weapons, decorated with flags and standard, pulled by four strongly built horses and piloted by a charioteer.[171] It goes on to say that a hard terrain was good for deploying the chariots.[172] The book also narrates the qualities a charioteer was expected to possess.[173]

The chariot used by the king was pulled by two or four horses, painted with the sketch of an elephant and decorated with flags of different colours.[174]

It is interesting that the *Mānasollāsa* also refers to a system of rewards (*pāritoṣikam*) granted to those, who excelled in destroying the enemy troops. The one who killed a charioteer in war was awarded one thousand coins by the king.[175]

KATHĀSARITASĀGARA

The *Kathāsaritasāgara*, composed during early medieval period, also provides valuable information on the constitution of chariot corps. Mayasura, when asked about the composition of his army by his son-in-law, stated the names of chariot warriors holding different ranks as follows:[176]

Ardharathī	Subāhu, Nirghāta, Muṣṭika, Gohara, Pralamba, etc.
Pūrṇarathī	Aṅkurin, Suviśāla, Daṇḍīn, Bhūṣaṇa, etc.
Dviguṇarathī	Prakampana, Darpita, Kumbhīra, etc.

Triguṇarathī	Vīrocana, Vīrasena, Yajñasena, etc.
Caturguṇarathī	Suśarmā, Bāhuśālin, Viśākha, Krodhana, etc.
Pañcaguṇarathī	Siṁhabhaṭṭa, Vyāghrabhaṭṭa and Śatrubhaṭṭa, etc.
Śaḍaguṇarathī	Ugravarmā
Saptaguṇarathī	Princes Viśoka, Sutantu, Sugama, etc.

Then followed the list of *mahārāthīs, mahārathī-nāyakas, atirathīs* and *adhipatis.*

DHANURVEDA

The *Dhanurveda* under the section *saṅgrāmavidhi* refes to the preparations made before joining a war. After performing necessary rituals the king mounted the chariot following his charioteer. He also yoked untiring horses to the chariot and equipped it with four bows and four hundred arrows, beside other weapons.[177] A chariot warrior fought with a bow made of bamboo[178] and knew how to hit a target while racing.[179] In the battle array chariots were deployed in the forefront.[180] The number of chariots in an *akṣauhiṇī* is stated to be 21,870[181] and that in an *mahākṣauhiṇī* 1,37,00,000.[182]

The evidence thus furnished by some Purāṇas, Nīti works, literary compositions, etc., which in no case can be dated earlier than the fourth century AD, suggests that chariots were very much in use in India during the Gupta period and even afterwards. If the chariots had faded out by this time, as believed by many, the chariot could not have been mentioned in such details in these works.

TEMPLE CHARIOTS[183]

Before ending, a word may be added here about temple chariots of India. Conceptualization of the chariot as a vehicle for gods goes back to the Vedic times. Gods then travelled in chariots to the *yajña* to partake *soma.* These chariots were made of wood and pulled by horses.

It appears that around fifth century AD, the tradition of representing chariot in architectural form evolved. This was perhaps done to facilitate the *darshan* of the diety by the devotees on specified days of *rathotsava.* A temple like structure was built on the wheels and the idol of the deity was mounted thereon for all to see. The chariot was pulled by the devotees through a specified route.

The tradition of chariot procession may have been popularized by the Mahāyāna Buddhists, who placed the golden image of Buddha in a chariot and carried it in procession through the city. This is indeed true because Fa-hien, the Chinese pilgrim, who visited India in fourth century AD, was himself a witness to one such ceremony during his sojourn at Khoṭan. It took place in the month of Vaiśākha on eighth lunar day. The car was shaped like a *stūpa* on four wheels with a height of 20 *hastas*. Golden idol of Buddha was placed at the centre and it was flanked by the idols of Boddhisattvas. Twenty chariots carrying idols of other divinities accompanied the procession.The festival of chariot started on the first day of the fourth month and ended on the fourteenth day. The chariots of the monasteries, which were decorated differently were taken out on specified days.[184]

About the same time the tradition became a part of Brāhmanic religious festivals. In the *Bhaviṣya Purāṇa*, there is a graphic description of a procession in which the idol of Brahmā, placed in a festival car, was taken around. The idol of Sāvitrī was placed on his right. The procession was organized on a full moon day in the month of Kārttīka. The chariot was pulled by the devotees and followed by learned Brāhmaṇas. No person, other than the *purohita* could mount the chariot.[185]

In the *Skanda Purāṇa*, the festival of cars in which idols of Vāsudeva, Balarāma and Subhadrā were carried in procession is described in detail. The dimensions of the car and its adornments are also mentioned.[186]

Meanwhile, literature came up on the production of temple chariots, their types, shapes and sizes. The *Mānasāra*, said to be the earliest treatise on architecture, devotes a chapter (43) to describe various kinds of chariots, the technique of constructing them and designing of the wheels. It is dated around sixth or seventh century. The other two important works related to the subject include *Mayamatam* (ch. 31) and *Aparājitapṛcchā* (ch. 195).

The *Mānasāra* mentions six types of chariots, viz., *nabhasvāna bhadraka, prabhañjana bhadraka, nivāta bhadraka, pavana bhadraka, indra bhadraka* and *anil bhadraka*. The *nabhasvāna* had four sides and was square in shape. *Prabhañjana* had six sides. Both displayed a *bhadra* symbol on the forefront. *Nivāta* and *pavana* displayed two and three *bhadra* symbols respectively. Indra had ten sides and one *bhadra* symbol. *Anila* had twelve sides.

Chariots have also been classified on the basis of the shape, circular, elliptical, square, etc. The chariot with four sides was called *nāgara* type, with eight sides *draviḍa* type, circular in form *vesara* type, with six sides *āndhra* type and with semi-circular back and two corners in the front *kaliṅga* type.

Another classification was based on the number of wheels. The chariots used in wars were provided with three wheels, while those used at sacrificial festivals were made of four wheels. But those used in other festivals were provided with five to ten wheels. The descriptions of chariot is the *Mayamatam* and the *Aparājitapṛcchā* generally follow the *Mānasāra*.

The *Viśvakarmā Śilpa*, a text attributed to Viśvakarmā, deals with chariots in detail. It classifies chariots in two categories – one used for travel purposes and the other used for temple festivals. It also describes the ritual of chariot consecration (*rathapratiṣṭhā vidhi*). *Viśvakarmīya Rathalakṣaṇam* is devoted to the chariots only. It is widely consulted by the makers of temple chariots in south India.

The said references to chariot types in architectural form find actual manifestation in early shrine models, particularly the Pāṇḍava *rathas* of Mamallapuram, named Draupadī Ratha, Arjuna Ratha, Bhīma Ratha, Dharmarāja Ratha and Nakula-Sahadeva Ratha. It is said that these Pāṇḍava *rathas* served as models for the later chariot-makers. The Sun Temple at Koṇārk has been designed as a massive chariot borne on twelve wheels.

Over the centuries chariots made of wood have become an integral part of the temple complex. These are taken out during the *rathotsavas*. Some famous *rathotsavas* are dedicated to Śṛṅgerī Śāradā, Mysore Chāmuṇḍeśvarī, Kukke Subramaṇya, Hampi Virūpākṣa, Gokarṇeśvara of Gokarṇa, Jagannātha of Purī, etc.

It seems that the tradition of *rathotsava* became popular after seventh century AD. Tamil works *Śilapaddikaram* and *Maṇimekaḷai* refer to *rathotsava* in some detail. The *Skanda Purāṇa* also refers to *rathayātrā*. But the mention of *deva-ratha* in the *Arthaśāstra* would carry back the antiquity of temple chariots by some five centuries.

NOTES

1. D. Schlingloff, *Studies in the Ajāntā Paintings*, p. 112.
2. P.C. Chakravarti, *The Art of War in Ancient India*, p. 26.
3. रथ कार्यस्तया युक्तः ईषादण्डकबन्धुरैः।
   सुचक्रोपस्करयुतः सर्वरूपः पताकराट्।। *Viṣṇudharmottara Purāṇa* (*VDP*), 2.183.10.
   त्रिवेणुसंततश्चैव वैयाघ्रपरिवारणः।
   किंकिणीजालनिर्घोषस्तथा वै सुप्रतिष्ठितः।। Ibid., 2.183.11.
   उपासङ्गैर्निषडैःश्च तुणीरैवेरकार्मुकैः।
   सर्वायुधैरुपेतश्च तथा वृक्षादनीयुतः।। Ibid., 2.183.12.
   तोयजीरकपिण्डीभिर्युक्तः शृङ्गैस्तथैव च।
   सन्नधः सारथिस्तत्र भवेद्राम प्रतोदवान।। Ibid., 2.183.13.
   रश्मिग्रहणविच्चापि तथा मण्डलकोविदः।
   तथाश्वहृदयज्ञश्च दूरदृष्टिः सुयंत्रितः।। Ibid., 2.183.14.
   स्वासनः कालविच्चैव बद्धासिः पाशसंयुतः।
   पार्ष्णिः स्यादायुधज्ञस्य तथैवायुधकोविदः।। Ibid., 2.183.15.
   चिरेण मूले यस्य स्यादवस्थानं महाभुजः।
   पताकावरणं यस्य आयुधानां समर्पणम्।। Ibid., 2.183.16.
   समायान्तं महाभागं तथा सारथिरक्षणम्।
   स्यातां धनुर्धरौ तत्र तथा च रथिनावुभौ।। Ibid., 2.183.17.
   योधौ धनुर्भृतां श्रेष्ठ सर्वायुधविशारदौ।
   तुरगाश्च सुसंनद्धा दान्ताः कार्यास्तथा रथे।। Ibid., 2.183.18.
   तुल्यवेगवलाश्चैव तथा लधुपरिच्छदा। Ibid., 2.183.19.
4. गजस्य देया रक्षार्थं चत्वारस्तु रथा द्विज।
   रथस्य चाश्वाश्चत्वारोऽश्वस्य तस्य च वर्मिणः। Ibid., 2.177.31-2.
5. त्रासने रिपुसैन्यानां रथकर्म तथोच्यते। Ibid., 2.177.47
6. Ibid., 1.261.66.
7. Ibid., 1.43.15-17; also 1.48.9-13.
8. चतुर्भिः कुञ्जरैर्युक्तं रथमास्थाय सत्वरा।
   तमन्वयौ धारादेवी देवं त्रिदशपूतितम्।। Ibid., 1.215-24.
9. चतुर्भिः कुञ्जरैर्युक्तं रथं दत्वा द्विजातये। Ibid., 3.312.15.
10. Ibid., 2.177.41.
11. दैत्यसैन्यातिस्तारवर्णनम्। सुरसैन्यविस्तारवर्णनम्। *Matsya Pu.*, 173-4.

12. रथं मातलिना क्लृप्तं देवराजस्य दुर्जयम्। Ibid., 148.81.
13. इत्येष एकचक्रेण सूर्य्यस्तूर्णं प्रसर्पति।
    हरिर्हरिद्भिः ह्रयते तुरङ्गमैः। Ibid., 126.39-40.
14. त्रिचक्रोभयतोऽश्वश्च विज्ञेयः शशिनो रथः।।
    अपाङ्गर्भसमुत्पन्नो रथः साश्वः ससारथिः।
    सहारैस्तै स्त्रिभिश्चक्रैर्युक्तः शुक्लैर्हयोत्तमैः।।
    संग्रहितारथे तस्मिन् श्वेतश्चक्षुः श्रवाश्चवै।
    अश्वास्तमेकवर्णास्ते वहन्ते शङ्खवर्चसः।।
    अजश्च त्रिपथश्चैव वृषो वाजी नरो हयः।
    अंशुमान् सप्तधातुश्च हंसो व्योम मृगस्तथा।। Ibid., 126.47-51.
15. Ibid., 127.1-13.
16. Ibid., 148.49-56; also 173.18. Maya and his commanders joined the war mounting chariots as follows:

Maya	Four-wheeled chariot pulled by 1,000 bears.
Tāraka	Eight-wheeled chariot pulled by 1,000 asses.
Grasan	Chariot pulled by 100 tigers.
Jambha	-do-
Kujambha	Chariot pulled by asses.
Mahiṣa	Chariot pulled by camels.
Kuñjara	Chariot pulled by horses.
Megha	Chariot pulled by large-sized leopards.
Kālanemi	Chariot pulled by large-sized elephants.
Nemi	Chariot pulled by mountain-sized elephants with four teeth.
Mathana	Chariot pulled by big black horses.
Tvaṣṭā	Chariot pulled by eighteen elephants.

    The mention of tigers and leopards as chariot pullars is simply wild imagination.
17. Ibid., 148.46-8.
18. Ibid., 173.
19. Ibid.
20. ततो रथै र्विप्रयुक्तैर्वारणैश्च प्रचोदितैः। Ibid., 175.4.
21. रथारथैर्निरुध्यन्ते पादाताश्च पदातिभिः।
    तेषां रथानां तुमुलः स शब्दः शब्दवाहिनाम्।।
    नभोनभश्चहि यथानभस्यै र्जलदस्वनैः।
    बभञ्जुस्तु रथान् केचित्केचित् सम्मर्दिता रथैः।।
    सम्बाधमन्ये सम्प्राप्य न शेकुश्चलितुं रथान्।
    अन्योन्यमन्ये समरे दोर्भ्यामुत्क्षिप्य दंशिताः।। Ibid., 177.31-33; also see 149, 153.
22. Ibid., 138.1-38.

23. तुरगरथप्रदानं। *Matsya Pu.*, 281.14.
कनकेभरथप्रदानम्। Ibid., 282.15; also 273.8; *Bhaviṣya Pu.* 180, 189.
24. निमित्तशकुनज्ञानी हयशिक्षाविशारदः।
हयायुर्वेदतत्त्वज्ञो भुवो भागविचक्षणः।।
बलबलज्ञो रथिनः स्थिरदृष्टिः प्रयंवदः।
शूरश्च कृतविद्यश्च सारथिः परिकीर्तितः।। Ibid., 215.20-1.
25. रथ्यासु राजमार्गेषु प्रसादेषु गृहेषु च।
दीपाश्चम्पकपुष्पाभा नाल्पस्नेहप्रदीपिताः।। Ibid., 139.19
26. रथ्यासु चन्द्रोदयभासितासु,
सुरेन्द्रमार्गेषु च विस्तृतेषु। Ibid., 139.38.
27. *Bhāgvata Pu.*, 10.76.13-15; This event is also described in Vana Parva of the *Mahābhārata*. Āraṇyaka, pp. 17-19 (BE).
28. धर्मविजानताऽऽयुष्मन् कृतमेतन्मया विभो।
सूतः कृच्छ्रगतं रक्षेद् रथिनं सारथिं रथी।।
एतद् विदित्वा तु भवान् मयापोवाहितोरणात्।
उपसृष्टः परेणेति मूर्छितो गदया हतः।। *Bhāgavata Pu.*, 10.76.32-33.
29. Ibid.
30. स्वलङ्कृतं श्यामतुरङ्गयोजितं
रथं मृगेन्द्रध्वजमाश्रितः पुरात्।
वृतो रथाश्वद्विपपत्तियुक्तया
स्वसेनया दिग्विजयाय निर्गतः।। Ibid., 1.16.11.

There are many more references to chariots in the *Bhāgavata Purāṇa*:

1. During the war between *devas* and *asuras* both sides fought mounting chariots. *Bhāg. Pu*, 8.10.8.
2. The dowry given during the marriage of Devakī included 1,800 chariots. Ibid., 10.1.31.
3. Pāṇḍava brothers went to see Bhīṣma in the battlefield mounting chariots. Ibid., 5.2.2.
4. Kṛṣṇa returned to Dvārakā from Hastināpur mounting a chariot. Ibid., 1.10.8.
5. To conquer heaven Bhṛgus performed Viśvajita *yajña*. From the *yajña-kuṇḍa*, a golden chariot pulled by green horses and mounted with *siṁha-dhvaja* emerged. Ibid., 8.15.5.

31. ससैन्ययानध्वजवाजिसारथी। Ibid., 10.50.21.
32. निघ्नत् रथान् कुंजरवाजिपत्तीन्। Ibid., 10.50.24.
33. हताश्वध्वजसूतनायकाः। Ibid., 10.50.25.
34. अथ देवस्य रुद्रस्य निर्मितो विश्वकर्मणा
सर्वलोकमयो दिव्यो रथो यत्नेन सादरम्।।
सर्वभूतमयश्चैव सौवर्णस्सर्वसंमतः।
रथांगं दक्षिणं सूर्यस्त्वाम्रं सोम एव च।।

दक्षिणं द्वादशारं हि शोडशारं तथोत्तरम्।
अरेषु तेषु विप्रेन्द्र आदित्या द्वादशैव तु।। *Śiva Pu.*, 2.5.8.5-7.
तस्य काष्ठाः स्मृता घोणाश्चाक्षदंडाः क्षणाश्च वै।
निमेषाश्चानुकर्षाश्च ईषाश्चानुलवाः स्मृताः।।
द्यौर्वरूथं रथस्यास्य स्वर्गमोक्षावुभौ ध्वजौ
युगान्तकोटितौ तस्य भ्रमकामदुधो स्मृतौ।। Ibid., 2.5.8.12-13; also see *Liṅga Pu.*, Pūrvārdha, 72.
In the Yuddha Khaṇḍa of *Śiva Purāṇa* there are many more references to chariot encounters. There is also a reference to *rathotsava* in this *Purāṇa* (1.9.23; also 5.51.67-70).

35. चिन्तामणिगृहेन्द्रस्य रक्षोभागेम्बुजाटवौ।
चक्रराजरथश्रेष्ठस्तिष्ठत्युन्नतविग्रहः।
नवभिः पर्वभिर्युक्तः सर्वरत्नमयाकृतिः।।
चतुर्योजनविस्तारो दशयोजनमुन्नतः।
यथोत्तरं ह्रासयुक्तः स्थूलतः कूवरोज्ज्वलः।।
चतुर्वेदमहाचक्रः पुरुषार्थमहाहयः।
तत्त्वैरु पचरद्भिश्च चामरैरभिमंडितः।।
पूर्वोक्तलक्षणैर्युक्तो मुक्ताच्छत्रेण शोभितः।
भंडासुरमहायुद्धे कृतसाहसिकक्रियः।।
वर्तते रथमूर्धन्यः श्रीदेव्यासनपाटितः।
Brahmāṇḍa Pu., 3.4.36.6-11; also 3.4.19.1.
गेयचक्ररथेन्द्रस्तु मंत्रिण्याः प्रान्त तिष्ठति। Ibid., 3.4.36.12; also 3.4.19.61.
वल्लभो दण्डनाथायाः किरिचक्रे महारथः। Ibid., 3.4.36.13; also 3.4.17.8.

36. तथा सह हयप्रायं सैन्यं हेषातरंगितम्।
व्यचरत्खुरकुद्दालविदारित महीतलम्।।
वनायुजाश्च कांबोजाः पारदाः सिंधुदेशजाः।
टंकणाः पर्वतीयाश्च पारसीकास्तथा परे।।
अजानेया घट्टधरा दरदाः काल वंदिजाः।
वाल्मीकयावनोद्भूता गान्धर्वाश्चाथ ये हयाः।।
प्राग्देशजाताः कैराता प्रांतदेशोद्भवस्तथा।
विनीताः साधु वोढारो वेगिनः स्थिरचेतसः।। *Brahmāṇḍa Pu*, 3.4.16.15-18.

37. *MB*, Bhīṣma, 90.3-5; *Rām*, Bāla, 6.22.

38. रथं तस्मै ददौ रुद्रः प्रीतः परमभास्वरम्।
असङ्गं काञ्चनं दिव्यमक्षयौ च महेषुधी।।
युक्तं मनोजवैरश्वैर्येन कन्यां समुद्वहन्।
स तेन रथमुख्येन जिगाय च ततो महीम्।।
ययातिर्युधि दुर्धर्षो देवदानवमानवैः।
पौरवाणां नृपाणां च सर्वेषां सोऽभवद्रथः।। *Vāyu Pu.* 93.18-20.

39. सोपासङ्गानुकर्षाणां सध्वजानां वरूथिनाम्।।
रथानां मेघघोषाणां सहस्राणि दशैव तु। Ibid., 96.121-22; also *Matsya Pu.*, 44.68.

40. *यौद्धो ध्वजो रथश्चैव प्रादुर्भवति मायया।* Ibid., 94.15.
41. *महारथाः।* Ibid., 94.48.
42. *स तेन रथमुख्येन जिगाय च ततो महीम्।* Ibid., 55.21 (S.S. edition)
43. *धनुंषि दश विस्तीर्ण श्रीमान् राजपथः स्मृतः।*
    नृवाजिरथनागानामसम्बाधः सुसंचरः।।
    धूनूंषि चैव चत्वारि शाखारथ्यास्तु तै कृताः।
    गृहरथ्योपरथ्याश्च द्विकाश्चाप्युपरथ्यकाः।। Ibid., 8.119-20 (S.S. edition).
44. *महिषासुरसेनानीश्चिक्षुराख्यो महासुरः।*
    युयुधेचामराश्चान्यश्चतुरंगबलान्वितः।।
    रथानामयुतैः षड्भिरुदग्राख्योमहासुरः।
    अयुध्यतायुतानां च सहस्रेण महाहनुः।।
    पंचाशखिश्च नियुतैरसिलोमा महासुरः।
    अयुतनां शतैः षडभिर्बाष्कलो युयुघे रणे।।
    गजवाजिसहस्रौघैरनेकैरुग्रदर्शनः।
    वृतोरथानांकोट्यांच युद्धे तस्मिन्नयुध्यत।।
    विडालाख्योमहादैत्यः पंचाशखिरथायुतैः।
    युयुधे संयुगे तत्ररथानां परिवारितः।।
    वृतः कालो रथानां च रणे पंचाशतायुतैः।
    युयुधे संयुगे तत्र तावद्भिः परिवारितः।। *Mārkaṇḍeya Pu.*, 79.40-45.
45. *कोटिकोटिसहस्रैस्तु रथानां दंतिनां तथा।* Ibid., 79.47.
46. *पतितैरथनागाश्वैरसुरैश्च वसुंधरा।*
    अगम्या साऽभवत्तत्र यत्राभूत्समहारणः।। Ibid., 79.66; also 80.5; 87.14.
47. *चिच्छेद सारथेश्चैव शिरः।* Ibid., 130.55.
48. Ibid., 130.55-8.
49. *रथगोऽश्वखरोष्ट्रांश्च।* Ibid., 130.62.
50. *हस्त्यश्वरथपादातं सैन्यं च परिकल्प्यताम्।* Ibid., 133.8.
51. *रथी च रथिना।* Ibid., 133.19.
52. *योग्यो नियुद्धनागाश्वस्यंदनाभ्यासतत्परः।* Ibid., 18.6.
53. *Agni Pu.*, 120.21-6.
54. *रथस्त्रिचक्रः सोमस्य कुन्दाभास्तस्य वाजिनः।* Ibid., 120.33.
55. Ibid., 120.36-7.
56. *हेमन्ते शिशिरे चैव रथवाजिसमाकुलम्।*
    चतुरङ्गबलोपेतां वसन्ते न शरन्मुखे।। Ibid., 228.6.
57. *पत्तिभुर्विषमा ज्ञेया रथाश्वानां तथा समा।* Ibid., 236.49.
58. *निस्थाणुवृक्षकेदारा रथमूभिरकर्दमा।*
    मर्दनीयतरुच्छेद्यव्रतती पङ्कवर्जिता। Ibid., 242.30.
59. Ibid., 242.45-8.
60. Ibid., 236.38-9.
61. *रथस्य चाश्वाश्चत्वारः।* Ibid., 236.38.
62. Ibid., 252.32.
63. *अभिन्नभेदनं मित्रसंधानं रथकर्म च।* Ibid., 242.25.

64. *त्रासनं रिपुसैन्यानां रथकर्म तथोच्यते।* Ibid., 236.48.
65. *अर्बुदे पर्वते सेनां कल्पयस्व सुदुर्घसम्।*
हस्त्यश्वकल्पितां भीमां रथपत्तिसमाकुलाम्।।
ततोऽसौ कल्पयामास चतुरंगां वरूथिनीम्।
पताकछत्रशवलां वदित्रावभूषिताम्।।
विमानप्रतिमाकारा रथास्तेन प्रकल्पिताः।
किंकिणीजाल संद्धटापताकाभिरलंकृताः।। Skanda Pu., 7.2.36.97-101.
66. *लक्षमेकं मतंगानां रथानां त्रिगुणं ततः।*
अश्वाः दशगुणा राजन्नसंख्याताः पदातयः।। Ibid., 7.2.36.103.
67. *गजरथतुरगाणां सम्बभूवातियुद्धम्।* Ibid., 5.16.2.
68. *Devībhāgavata Pu.*, 3.15.43.
69. *शतद्वयं चाश्वयुजां रथानां*
सुभूषितं चापि शरौधसंयुतम्।
ददौ नृपेन्द्रस्तु सुदर्शनाय
सपूजितं पारिबर्हं विवाहे।। Ibid., 3.22.15.
70. *Ādi Purāṇa*, 26.69-72.
71. *पुरः पादातमाश्वीयं रथकड्या च हस्तिकम्।* Ibid., 26.77.
72. *अस्त्रैर्व्यस्त्रैश्च शस्त्रैश्च शिरस्त्रैः सतनुत्रकैः।*
दधुर्जयनशालानां लीलां रथ्याः सुसम्भृताः।। Ibid., 31.72.
73. *रथकट्यापरिक्षेपो बलस्योभयपक्षयोः।* Ibid., 36.4.
पादातहस्तिकाश्वीयरथकट्यापरिच्छदैः। Ibid., 36.18.
74. Ibid., 44.177-92.
75. *अधिरुह्य रथं जम्बूर्नगरद्वारमाययौ।* Pariśiṣṭaparvan, 2.99.
गच्छतो मेऽध्वनानेन शिलोपरि पतेद्यदि।
तदास्मि नाहं न रथो न रथ्या न च सारथिः।। Ibid., 2.107.
76. *Mahāpurāṇa* (Adi) 8; also
ततः करीन्द्रैस्तुरगैः पत्तिभिश्चद्यतायुधैः नृपाजिरमभूद्
रुद्धं स्यन्दनैश्च समन्ततः। Ibid., 8.136.
रथानां वारणाञ्च केतवो। Ibid., 8.139.
77. *तरस्विमिर्वपुर्मेधावयः सत्वगुणान्वितैः।*
तुरङ्गमैस्तुरुष्का द्यैः विभुमाराधयन् परे।।
केचित्काम्बोजबाह्लीकतैतिलारट्टसैन्धवैः।
वानायुजैः सगान्धारैः बापेयैरपि बाजिभिः।।
कुलोपकुलम्भूतैः नानादिग्देशचारिभिः।
आजानेयैः समग्राङ्गैः प्रभुमैक्षन्त पार्थिवाः।।
Mahā Purāṇa, 30.106-8; also compare:
वनायुजाश्च कांबोजाः पारदाः सिंधुदेशजाः।
टंकणाः पर्वतीयाश्च पारसीकास्तथा परे।।
अजानेया घट्टधरा दरदाः काल वंदिजाः।
वाल्मीकयावनोद्भूता गान्धर्वाश्चाथ ये हयाः।।

प्राग्देशजाताः कैराता प्रांतदेशोद्भवस्तथा।
विनीताः साधु वोढारो वेगिनः स्थिरचेतसः। *Brahmāṇḍa Pu.*, 3.4.17.16-18.

78. *Vāyu Pu.*, chaps., 51-2; also *Agni Pu.*, 120.24-38.
79. *संस्थितेनैकचक्रेण पञ्चारेण त्रिनाभिना।* Ibid., 51.54.
80. *स तस्य ब्राह्मणा सृष्टो रथो ह्यर्थवशेन तु।*
असंगः काञ्चनो दिव्यो युक्तः परमगैर्हयैः।। Ibid., 51.57.
अहस्तु नाभिः सूर्यस्य एकचक्रः स वै स्मृतः। Ibid., 51.60.
रथनीडः स्मृतो ह्यब्दस्त्वयने कूबरावुभौ।
मुहूर्ता बन्धुरास्तस्य शम्या तस्य कलाः स्मृताः।।
तस्य काष्ठाः स्मृता घाणा ईशादण्डः क्षणास्तु वै।
निमेषाश्चानुकर्षोऽस्य ईषा चास्य लवाः स्मृताः।।
रात्रिर्वरूथो धर्मोऽस्य ध्वज ऊर्ध्वः समुच्छ्रितः।
युगाक्षकोटी ते तस्य अर्थकामावुभौ स्मृतौ।। Ibid., 51.61-3; also see 51.64-70, 52.42-3, 54-69; also *Matsya Pu.*, 125.37-40; *Viṣṇu Pu.*, 2.10-11; *Bhaviṣya Pu.*, 51.60.
81. The names of these horses were *गायत्री, त्रिष्टुभ, अनुष्टुभ, जगती, पंक्ति, वृहती,* and *उष्णिक।* Ibid., 51.64-5; *Matsya Pu.*, 125.46-7.
82. *हरिर्हरिद्भिर्हियते:।* Ibid., 52.42; The chariot of Sun is described in detail in *Kūrma Pu.* Pūrvārdha, 41-2; *Matsya Pu.*, 126.39-46.
83. C. Nandgopal, and V. Iyenger, *Temple Treasures,* vol. III, p. 38.
84. *त्रिणा भिन्ना तु चक्रेण पंचारेण समन्वितः।* *Liṅga Pu., Pūrvārdha*, 55; also see *Kūrma Pu., Pūrvārdha*, 41.
85. *स्वस्थां प्रतिमां कृत्वाकुर्यात्स्थान प्रदिक्षिणाम्।*
एवं तु क्रियमाणायां यात्रायां प्रतिवत्सरम्।। *Sāmba Pu.*, 34.10.
86. *Bhavisya Pu.*, 55.59-61.
87. *त्रिचक्रोभयपार्श्वस्थो विज्ञेयः शशिनो रथः।*
अपां गर्भसमुत्पन्नो रथः साश्वः ससारथिः।।
शतारैश्च त्रिभिश्चक्रैर्युक्तः शुक्लैर्हयोत्तमैः।।
दशभिस्तु कृशैर्दिव्यैसङ्गैस्तैर्मनोजवैः।
सकृद्युक्ते रथे तस्मिन् वहन्ते चाऽऽयुगक्षयात्।। Ibid., 52.50-1.
Ten horses included *ययु, त्रिमना, वृष, राजीबल, हय, वाम, तुरण्य, हंस, व्योम, मृग।* Ibid., 52.53-4; also see *Brahmāṇḍa Pu.*, 1.2.23. 51-7; *Kūrma Pu.*, 41.28-30; *Matsya Pu.* 126.47-5; *Viṣṇu Pu.*, 2.12.1.
88. *तोयतेजोमयः शुभः सोमपुत्रस्य वै रथः।*
युक्तो हयैः पिशङ्गैस्तु अष्टाभिर्वातरंहसैः।।
सवरूथः सानुकर्षः सूतो दिव्यो रथो महान्।
सोपासङ्गपताकास्तु सध्वजो मेधसंनिभः।। Ibid., 52.72-3; *Viṣṇu Pu.*, 2.12.16; *Matsya Pu.*, 127.1-3; In the *Matsya Purāṇa* the number of horses is said to be ten (*दशभिर्वातहंसैः*)।
89. *भार्गवस्य रथः श्रीमांस्तेजसा सूर्यसंन्निभः।*

पृथ्वीसम्भवैर्युक्तैः नानावर्णैर्हयोत्तमैः।।
श्वेतः पिशङ्गः सारंगो नीलः पीतो विलोहितः।
कृष्णश्च हरितश्चैव पृषतः पृश्णिरेव च।
दशभिस्तैर्महाभागैरकृशैर्वातवेगितैः। Ibid, 52.75-6; also see *Viṣṇu Pu.*, 2.12.17.

90. अष्टाश्वः काञ्चनः श्रीमान् सोमस्यापि रथोऽभवत्।
असंगैर्लोहितैरश्वैः सर्वगैरग्निसम्भवैः।
सर्पतेऽसौ कुमारो वै ऋजुवक्रानुचक्रगः। *Vāyu Pu.*, 52.77; *Matsya Pu.*, 127.4-5; *Viṣṇu Pu.*, 2.12.18.
91. तवस्त्वागिरसो विद्वान् देवाचार्यो बृहस्पतिः।
शोणैरश्वैः कांचनेन स्यन्दनेन प्रसर्पति।। Ibid., 52.78.
युक्तस्तु वाजिभिर्दिव्यैरष्टाभिर्वातसंमितैः। Ibid., 52.79; *Viṣṇu Pu.*, 2.12.19.
92. गोराश्वेन तु रौप्येण स्यन्दनेन विसर्पति। *Matsya Pu.*, 127.6.
93. ततः शनैश्चरोऽप्यश्वैः शबलैर्व्योमसम्भवैः।
कृष्णायसं समारुह्य स्यन्दनं याति वै शनैः।। *Vāyu Pu.*, 52.80; *Matsya Pu.*, 127.7-8; *Viṣṇu Pu.*, 2.12.20.
94. स्वर्भानोस्तु तथैवाश्वाः कृष्णा ह्यष्टो मनोजवाः।
रथं तमोमयं तस्य सकृद्युक्ता वहन्त्युत।। Ibid., 52.81; *Viṣṇu Pu.*, 2.12.21.
95. अथ केतुरथस्याश्वा अष्टाष्टौ वातरंहसः।
पलालधूमसङ्काशाः शवला रासभारुणाः।। Ibid., 52.83; *Matsya Pu.*, 127.10-11; *Viṣṇu Pu.*, 2.12.23.
96. R.L. Mitra, *Antiquities of Orissa*, vol. 1, p. 131.
97. स्निग्धगंभीरनिर्घोषमेकं स्यन्दनमाश्रितो। *Raghuvaṁśa*, 1.36; aso 1.39-40; There are many more references to chariot in the *Raghuvaṁśa.*
98. Ibid., 5.28.
99. Ibid., 5.49.
100. *Abhijñānaśākuntalam*, Act 1.
101. *Vikramorvaśīyam*, Act 1.
102. Ibid., Act. 1.5.
103. Ibid., 3.42-43.
104. *Raghuvaṁśa*, 4.30.
105. Ibid., 5.49; 7.36, 52.
106. *Abhijñānaśākuntalam*, Act 7.
107. अजयदेकरथेन स मेदनी। *Raghuvaṁśa*, 9.10.
108. Ibid., 7.52.
109. *Pratijñāyogandharāyaṇa*, p. 44; also see pp. 88, 124.
110. *Pañcarātra*, pp. 186-9, 224-5, 228.
111. *Urubhaṅga*, pp. 338, 340.
112. रवितुरंगसमानं साधनं राजलक्ष्म्याः।
सकलनृपति मान्यकाम्बोजजातम्।। *Karṇābhāra*, 1.19.
113. *Kāmandaka*, 18.24; *ṣaḍāṅgabala* comprised infantry, cavalry, chariot corps, elephant corps, treasury and counsel.

114. Ibid., 15.48.
115. Ibid., 19.24-5.
116. Ibid., 19.36-7.
117. Ibid., 18.46.
118. Ibid., 19.11-12.
119. Ibid., 19.13.
120. *Śukranīti*, 1.226; also *रथाश्वगजगं त्रिधा।* *Śukranīti*, 4.7.3; also 4.7.25; 4.7.22; The royal chariot is described as follows:
लोहसारमयश्चक्रसुगमो मञ्चकासनः।।
स्वान्दोलायितरूढस्तु मध्यमासनसारथिः।
शस्त्रास्त्रसन्धायुदर इष्टच्छायो मनोरमः।।
एवंविधो रथो राज्ञा रक्ष्यो नित्यं सदश्वकः। Ibid., 4.7.30-2.
121. *गजाश्वरथयानं तु प्रातः सायं सदाभ्यसेत्।* Ibid., 1.331.
122. *अश्वानां हृदयं वेति जातिवर्णभ्रमैर्गुणान्।।*
गतिं शिक्षां चिकित्सां च सत्त्वं सारं रुजं तथा।
हिताहितं पोषणं च मानं यानं दतो बयः।।
शूरश्च व्यूहवित्प्राज्ञः कार्याऽश्वाधिपतिश्च सः।
एभिर्गुणैश्च संयुक्तो धुर्यान्युग्यांश्च वेत्ति यः।।
रथस्य सारं गमनं भ्रमणं परिवर्तनम्।
समापतत्सु शस्त्रास्त्रलक्ष्यसंधाननाशकः।।
रथगत्या स रथयो हयसंयोगगुप्तिवित्। Ibid., 2.130-4.
123. *नौकारथादियानानां कृतिज्ञानं कला स्मृता।* Ibid., 4.3.87.
124. *रथकारिणः।* Ibid., 2.199.
125. *गजाश्वरथगत्या तु युद्धसंयोजनं कला।* Ibid., 4.3.82.
126. *रथेन च रथो योज्यः।* Ibid., 4.7.357.
127. Ibid., 1.226-7.
128. *एकाश्वरथवाहनौ।* Ibid., 5.82.
129. *द्विअश्वयानगः।* Ibid., 5.83.
130. *यानैश्च चतुरश्वगः।* Ibid., 5.84.
131. *याने चक्राणि युद्धार्थे शताङ्गः स्यदनो रथः।* *Amarakośa*, 2.8.51.
132. *असौ पुष्यरथचक्रयानं न समराय यत्।* Ibid.
133. *कर्णीरथः प्रवहणं डयनं च समं त्रयम्।* Ibid., 2.8.52.
133. *उभौ तु द्वैपवैयाघ्रौ द्वीपिचर्मावृते रथे।* Ibid., 2.8.53.
134. *पाण्डुकम्बलसंवीतः स्यन्दनः पाण्डुकम्बली।* Ibid., 2.8.54.
135. *रथे काम्बलवास्त्राद्याः कम्बलादिभिरावृते।* Ibid.
136. Ibid., 2.8.55-7.
137. Ibid., 2.10.9.
138. Ibid., 2.10.4.
140. *तिनिशे स्यन्दनो नेमी रथद्रुरतिमुक्तकः।। वञ्जुलश्चित्रकृचा।* Ibid., 2.4.26-7.
141. Ibid., 2.8.59-60.
142. *रथिनः स्यन्दनारोहाः* Ibid., 2.8.60.

143. *रथ्यो वोढा रथस्य यः।* Ibid., 2.8.46.
144. *युद्धार्थे चक्रवद्याने शताङ्गः स्यन्दनो रथः।*
संक्रीड़ार्थः पुष्परथो देवार्थस्तु मरुद्रथः।।
योग्यो रथो वैनायिकोऽध्वरथः परिघातिकः।
कर्णी रथः प्रहरणं उडयनं रथगर्भकः।। Hemacandra quoted by R.L. Mitra, *Indo-Āryans*, p. 347.
145. *Mānasāra*, chap. 43
146. *Aparājitapṛcchā*, 70.23; 86.33.
147. *शनैरनीयन्त रयात्पतन्तो रथाः क्षितिं हस्तिनरवादखेदैः।*
सयत्नसूताऽयतरश्मिभुध्नग्रीवाग्रसंसक्तयुगैस्तुरङ्गैः।। *Śiśupālavadha*, 3.68; also see 5.8, 5.20, 12.8, 12.15, 13.4, 20.4.
148. *रथौधैः।* Ibid., 3.27.
149. Ibid., 5.9.
150. *द्रुतद्रवद्रथचरणक्षतक्षमातलोल्लतदबहुलरजोबगुण्ठितम्।*
युगक्षयक्षणनिरवग्रहे जगत्पयोनिधेर्जल इव मग्नमावभौ।। Ibid., 17.60.
151. *रथमण्डलस्वनैः* Ibid., 12.15.
152. *स्रस्ताऽङ्गसन्धौ विगताऽक्षपाटवे रुजा निकामं विकलीकृते रथे।*
आपतेन तक्ष्णा भिषजेव तत्क्षणं प्रचक्रमे लङ्घनपूर्वकः क्रमः।। Ibid., 12.25.
153. Ibid., 18.2, 18.3, 18.7, etc.
154. *रराज सम्पादकमिष्टसिद्धेः सर्वासुदिक्ष्वप्रतिसिद्धमार्गम्।*
महारथः पुष्यरथं रथाङ्गी क्षिप्रं क्षपानाथ इवाधिरुढः।। Ibid., 3.22.
155. Ibid., 3.23.
156. Ibid., 14.18.
157. *युवराजस्य वर्वाणां पंचसाहस्रकी भृतिः।*
सर्वसेनाप्रणेत्रे च चतुसाहस्रकी च सा।।
भृतिचातिरथे देया वर्वाणाम् त्रिंसहस्रकम्।
महारथाय सहस्रद्वयम् राज्ञाधिमासकम्।।
वेतनम् रथिकायाथ साहस्रम् गजयोधिने।
दद्यात् अर्धरथायाथ वेतनं शतपंचकम्।।
एकस्मै रथिकायाथ तादृशे गजसादिने।
निष्काणाम् त्रिशतम् दद्यात यतस्तौ तत्कुटुम्बिनौ।। *Nītiprakāśikā*, 7.33-6.
158. *गजयन्तुस्सरथेश्च ध्वजिने चक्रपाय च।*
मासि मासि तु वर्वाणाम् दशपंच तु वेतनम्।। Ibid., 7.40-2
159. *द्वितीय सैन्यदायित्वं गजसंरक्षणं तथा।*
भिन्नसन्धानकारित्वं दूरे शत्रुनिवारणम्।।
शत्रूणामाभिरोधश्च व्यूहद्वारे व्यवस्थितः।
महाघोषविविश्चेति रथकर्म प्रचक्षते।। Ibid., 6.62-3.
160. *प्रासादवेश्मपरिखाद्वाररथ्यासभादिषु।*
मार्गाश्च निर्गमा येषां सीमक्षेत्रान्तराणि च।। *Samarāṅgaṇa-Sūtradhāra*, 9.32.

161. बलस्य चतुरङ्गस्य पौराणां पार्थिवस्य च।
असम्बाधसमश्चैष कार्योऽयं काश्मशर्करः।। Ibid., 10.8.

162. महारथ्याद्वयं कार्यं तदुपान्तस्थवंशयोः।
तद् द्वादश दशाष्टौ स्यात् करान् ज्येष्ठादिकं त्रिषु।। Ibid., 10.9.
पदमध्यगतं कार्यं यानमार्गचतुष्टयम्।
ज्येष्ठादिषु पुरेष्वेषु तत्पद्यं च चतुःकरम्।। Ibid., 10.10.
उपरथ्यामहामार्गस्यार्धं वा द्विशयाधिकम्।
शेषा रथ्यास्तदर्धेन विघातव्याः प्रमाणतः।। Ibid., 10.11.
महारथ्याप्रमाणेनतद्भूमेर्वाह्यतस्ततः।
व्यासखातान्तरैः सार्धं विधेयं परिखात्रयम्।। Ibid., 10.17.
राजमार्गमहारथ्यासंश्रितानि चतुर्दिशम्।
त्रीणि त्रीणि विधेयानि पुरे द्वाराणि तद्विदा।। Ibid., 10.35.
There are many more references to *rathyā*. महारथ्याश्रयं द्वारम्। Ibid., 10.37; रथ्याचत्वरवर्त्मभिः। 3.15; also 9.32, 39.39, 53; 45.3.

163 रथेषु कौशलं येषां येषां स्यादायुधेषु च।
वारुण्यां दिशि तान् सर्वान् पुरस्य विनिवेशयेत्।। Ibid., 10.92.

164. कार्या तैतस्नते शाला रथानां दन्तिनां तथा। Ibid., 15.23.

165. भूमिभागवशात् कल्पं महारथ्योभयान्वितम्।
शिविरस्य तु चत्वारि कुर्याद् द्वाराणि यत्नतः।। Ibid., 45.32.
रथ्या सार्धा तु सेनायाः पुररथ्याप्रमाणतः।
मित्रे स्थानं नरपतेः कार्यं पृथ्वीधरेऽपि वा।। Ibid., 45.33.

166. रथिकाशिखाग्रचक्रं भ्रमचक्रारकविनियोजितं कार्यम्।
यष्टिचतुष्टयमस्मिंस्तिर्यक् चक्रद्वयोपेतम्।।
ऊर्ध्वं द्वितीयभूमेस्तृतीयभूमेस्थान्तरे कुर्यात्।
नियतं रथिकायष्टिभ्रमसंलग्नानि यन्त्राणि।।
आसनाधारयष्टीनां रथिकाचक्रयोजितान्।
अधः समान्तरान् कुर्याच्चतुरः परिवर्तकान्।। Ibid., 31.211-13. There are many more details on chariot construction in this chapter.

167. For details see *Samarāṅgaṇa-Sūtradhāra*, chapters 54 to 58.

168. युद्धार्थे चक्रवद्याने शताङ्गः स्यन्दनो रथः।
स क्रीडार्थः पुष्परथो देवार्थस्तु मरुद्रथः।
योग्यारथो वैनयिकोऽश्वरथः परियानिकः।।
कर्णीरथः प्रवहणं डयनं रथगर्भकः। *Abhidhāna Cintāmaṇi*, 3.415-17.

169. Ibid., 13.419.

170. Ibid., 3.420-5.

171. अनेकायुधसङ्कीर्णाः पताकाध्वजराजिताः।
चतुर्मिस्तुरगैर्युक्ता दृढाः सारथिनोदिताः।। *Mānasollāsa*, 2.679-80.

172. स्थिरा चक्रसभा धात्री रथव्यूहे प्रशस्यते।। Ibid., 2.1174.

173. आयुर्वेदविदश्वानं निमित्त-शकुनादिवित्।
शिक्षावेदीतुरङ्गाणां भूमिभागविशारदः।।

बलाबलरथाभिज्ञः प्रियवाक् प्रियदर्शिनः।
कृतविद्यश्च शूरश्च सारथी पार्थिवोचितः।। Ibid., 2.132-3.

174. अश्वैश्चतुर्भिर्द्वाभ्यां वा युक्तश्चक्रयुगान्वितः।
मत्तवारणकैर्भव्यैश्चित्रितैः परिशोभितः।।
नानावर्णपताकाभिर्भ्राजितः सुदृढाक्षकः।
उत्तमः स्यन्दनो याने नृपाणामेव निर्मितः।। Ibid., 16.55-6.

175. सहस्रं प्रतिदास्यामि स्यन्दनस्य विघातिनाम्। Ibid., 20.1166.

176. *Kathāsaritasāgara*, 8.4.12-32.

177. पूर्वं सारथिमारोप्य रथे सज्जेत्ततः स्वयम्।
योजयेद्वाजिनः शुद्धान्सुसंतुष्टाञ्जितश्रमान्।।
रथे च धारयेत् भद्रं कार्मुकाणां चतुष्टयम्।
चतुःशतानि वाणानां तूणीरे च प्रयोजयेत्।।
Śāraṅgadharapaddhati, 80.189-91.

178. रथिनां च पदातीनां वांशं चापं प्रकीर्तितम्। Ibid., 80.46.

179. धावता वै श्रमः कार्यो लक्ष्यं हन्तु सुनिश्चितम्। Ibid., 80.157.

180. मुखे रथा गजाः पृष्टे तत्पृष्टे पदातयः। Ibid., 80.208.

181. Ibid., 80.197.

182. Ibid., 80.203.

183. For details please refer to *Temple Treasures* (vol. III), by Choodamani Nandgopal and Vatsala Iyenger.

184. J. Legge, *The Travels of Fa-hien*, pp. 18-19.

185. तस्माद्रथेन देवस्य यात्रा कार्या विधानतः।
प्रजानामिह शान्त्यर्थं प्रतिसंवत्सरं सदा।।
काञ्चनो वाथ रौप्यो वा दृढ़दारुमयोऽपि वा।
दृढ़ाक्षयुगचक्रश्च रथः कार्यः सुयन्त्रितः।।
तस्मिन् रथवरे श्रेष्ठे कल्पिते सुमनोरमे।
आरोष्यं प्रतिमां यत्नायोजयेद्वाजिनः शुभान्। *Bhaviṣya Pu.*, 1.55.59-61; also *Sāmba Pu.*, 34.10, 22-3.

186. *Skanda Purāṇa*, 2.2.25.8-12; The width of the festival car of Vāsudeva is said to be 16 *hastas* (7.3 m) of Balarāma 14 *hastas* (6.384 m) and of Subhadrā 12 *hastas* (5.472 m). The number of wheels in their cars was 16, 14 and 12 respectively. The cars were nicely decked:
स्वक्षं सुचक्रं सुस्तम्भं सुविस्तीर्णं सुतोरणम्।
सुध्वजं सुपताकं च नानाचित्रमनोहरम्।।
विचित्रबन्ध मिथुनपुत्तलीबलयान्वितम्।
अर्द्ध हाटकनिर्व्यूढं साक्षाद्रविरयोपमम्।
मेघ गंभीरनिर्घोषं दष्ट्वा कर्षगुणैर्युतम्।। *Skanda Purāṇa*, 2.2.25-19-21.
It may be mentioned here that the festival-car of Vāsudeva at Purī is 13.7 m high, 10.67 m in square and is raised on 16 wheels of 2.13 m diameter each.

CHAPTER 8

The Eclipse

It is likely that with the emergence of the horse as the most effective vehicle of war in the wake of foreign invasions, the chariot lost its primacy in the Indian battlefield, around the beginning of the Christian era. However, it did not fade out, as believed widely, and continued to be a part of the Indian armies for many more centuries. The Junāgaḍha inscription of Rudradāman, the Śaka ruler, written in 151 AD, refers to the chariot arm.[1] It clearly suggests that the foreign invaders, who ruled over India around the beginning of the Christian era also patronized this arm.

The chariot formed a part of the armies of India in seventh century AD is incontrovertibly stated by Chinese traveller Hiuen Tsang:

> There are four divisions of the army viz. (1) the infantry (2) the cavalry (3) the chariot (4) the elephants. . . . A leader in a car gives the command while two attendants on the right and left drive his chariot, which is drawn by four horses, abreast. The general of the soldiers remains in his chariot; he is surrounded by a file of guards, who keep close to his chariot wheels.[2]

Further, the *skandhāvāra* (military camp) of Bhojadeva I (706 AD) included chariots besides elephants, horses, boats and foot-soldiers.[3] Samnad Plate belonging to the Rāṣṭrakūṭa ruler Dantivarman II (735-56) records that he defeated a large Karṇāṭaka army with his army of chariots and horses.[4]

A terracotta plaque discovered from Ahicchatra, displayed in the National Museum, depicts two chariot warriors engaged in a battle (Figure 8.1). The plaque dated around seventh century AD suggests that the chariot still formed a part of Indian armies. The representation of chariots in the sculptures at Halebiḍ

Figure 8.1: Chariot Encounter – Ahicchatra

(Figures 8.2, 8.3), Pahārpur (Figure 8.4) and Puṣpagiri (Figure 8.5) also supports this. In the Nālandā Copper Plate of Dharmapāla, chariot is included in the five-fold division of his army.[5] The Khaira Plate of Yaśahkaraṇa (1076 AD) proves that chariots formed a part of Kalchuri army as late as the eleventh century.[6]

Further, the Purāṇic evidence on the chariot arm, covering the period from fourth to eighth century AD is overriding and does not admit any other conclusion. It would, therefore, be difficult to agree with Basham that 'chariot as a vehicle of war disappeared soon after the commencement of the Christian era'.[7]

However, there is substantial evidence on the diminishing presence of the chariot in the Indian battlefield. There is no epigraphic and numismatic evidence that the Gupta rulers patronized the chariot arm. The *Harṣacarita* of Bāṇa Bhaṭṭa is also silent about the use of the chariot by his patron king. Surprisingly, Muhammad-bin-Qasim, the Arab invader, did not encounter any war chariots during his Indian conquest.

Figure 8.2: Chariot Battle – Kedāreśvara Temple, Halebiḍ

Figure 8.3: Chariot Battle – Hoysaleśvara Temple Halebiḍ

Figure 8.4: Chariot Warrior – Pahārpur

Figure 8.5: Chariot Battle – Puṣpagiri

Perhaps, to this diminishing presence foreign invaders contributed the most. They introduced new tactics of war in the Indian battlefield, hitherto unknown to Indians. Their mobile cavalry with sharp manoeuvres could always surprise and deceive the Indian defenders, who banked on their chariots. Moreover, the utility of chariot arm was also compromised as it could fight only in good weather conditions and even terrains.

P.C. Chakravarty, who has done some in-depth study on the subject has concluded that chariot disappeared from the Indian military system about the eighth century.[8] He goes on to say that 'there is not an iota of evidence to prove that they (chariots) were employed by the Gupta's and it is certain that they were not used in the empire of Harṣa'.[9] Macdonell seems to support him while observing that, 'the use of chariot both for war and for racing gradually died out in Hindustan, partly perhaps owing to the enervating influence of the climate and partly to the scarcity of horses, which had to be brought from the region of Indus'.[10]

Though the chariot might have lost its primacy over the centuries it would not be correct to say that it 'disappeared' or 'died out' about the eight century. There is substantial literary, epigraphic and art evidence to negate this.

Some irrefutable evidence on the presence of chariot even after eight century is provided by the *Mānasāra* and related works on architecture. The *Mānasāra* clearly says that, while planning a township the houses of the charioteers (*sūta-kāra*) should be built in the south and south-west.[11] Further, the chariot warriors were to be lodged in *caturmukha* type of houses alongside those who fought mounted on horses and elephants.[12] In the planning, due care was taken to provide for large streets called *rathyā* for the smooth movement of chariots.[13] Chariot was considered as a fast conveyance[14] and men of resources maintained chariots along with horses, elephants and camels.[15] it is also notable that *Mānasāra* has devoted a complete chapter to describe the chariots and the technique of their construction. Following the *Mānsāra*, the *Mayamatam*, a medieval text while describing the chariots, says that these could be mounted on occasions like the attainment of *chakravartin* status by a king, waging war, annual festival, worship of gods, etc. This wide use of the chariot is further corroborated by the *Samarāṅgaṇa Sūtradhāra* of Bhoja and *Nīti Prakāśikā* of Vaiśampāyana as detailed above.

As a token of regard to the chariot, the festival of *rathasaptamī* also came into vogue about this time. It was celebrated on the seventh *tithi* of the bright fortnight of Māgha in the honour of God Sūrya, who is said to have revealed his form to gods on this day. It was considered a pious deed to gift a chariot on this day. One Kalchuri king Daddha is reported to have made the gift of a chariot pulled by elephant on this day in 679 AD.[16] Perhaps, the horse-driven chariots were widely gifted. Indian love for chariots also found expression in various art forms and festivities. The Pāṇḍava *rathas* at Mahābalipuram built by Pallavas are the architectural representations of the actual chariots. The massive *rathas* with elaborate carvings commonly depicted in south Indian temples are said to have been inspired by the Mahābalipuram models. The conceptualization of chariot as a vehicle of gods is epitomized in many medieval temples. The Śārangapāṇi Temple at Kumbhakoṇam and the Sun Temple at Koṇārka are modelled in the form of chariots.

In the later centuries, the chariots made of wood became an integral part of Indian temples, to serve as the vehicle of the presiding diety. The wood chariot at the Virūpākṣa Temple at Hampī, which was certainly a later addition, was bigger than the original stone chariot. It was pulled through the streets outside the temple by the devotees on festive occasions. The *Rathatsovas* are celebrated with great grandeur at Veṅkaṭaramaṇa Temple at Mangalore, Śrī Kṛṣṇa Temple at Uḍupī, Śrī Śaradā *pīṭham* at Śṛṅgerī and many more temples in south India. But the *Rathayātrā* of Lord Jagannāth at Purī epitomizes this ancient tradition in the most gigantic scale even today. The temple was built by Gaṅga rulers in twelfth century AD.

These temple chariots were indeed considered sacred and great care was taken in constructing them. The *Mānasāra, Mayamatam, Viśvakarmīya Rathalakṣanam, Aparājitapṛcchā, Samarāṅgaṇa-sūtradhāra*, etc., describe the characteristics of these temple chariots in full detail.

It may not be out of place to mention here that the celebration of car fastival had started in India earlier than fourth century AD. Fa-hien has referred to a car celebration held annually at Pāṭaliputra. During this festival richly adorned images of Buddha and *bodhisatvas* were carried in a procession. It was perhaps an old Brāhmaṇic tradition which was copied by the Buddhists.

Thus, Indian art and literature is replete with references to chariot in various forms. Though Indians do not appear to have used it as a war machine after the arrival of Turks it continues to survive as a vehicle of the gods even today.

CHARIOT: A POLITICAL TOOL

It may also be mentioned here that more recently the chariot has been rediscovered by Indian politicians. It is being used as a vehicle to fight political opponents by conducting *ratha-yātrās*. In these *yātrās* leaders travel across the country mounting motorized vehicles named Vijaya Ratha, Rāma Ratha, etc. fitted with loud-speakers. Like Arjuna hurling arrows from his Gāṇḍīva, these leaders hurl abuses by means of loudspeakers on their adversaries to destroy them in the electoral arena.

In recent history it was the Telugu leader N.T. Rama Rao who effectively used a motor driven chariot called Caitanya Ratha to fight his political battle. He converted his old Chevrolet car into a *ratha* and travelled through the Āndhra territory in 1982. This *ratha-yātrā* enabled him to vanquish his adversaries and win the election in 1983.

In 1989, taking a leaf out of N.T. Rama Rao's book, Jat leader of Haryana Devi Lal also conducted a *yātrā* in his Vijaya Ratha to usurp PM's chair in Delhi. His chariot, a luxury car, was accompanied by as many as 18 video chariots or propaganda vans, which screened his speeches during the march.

The Rāma Ratha of L.K. Advani, however, proved to be the most powerful vehicle in his conquest of Ayodhyā. It was a mini-truck shaped like a chariot piloted by two charioteers named Narendra Modi and Pramod Mahajan. The Rāma Ratha travelled some 10,000 km., from Somnāth to Ayodhyā to garner support for building a Rāma Temple at the latter site. Advani conducted some more *ratha-yātrās* to vanquish his adversaries.

The Namo Ratha was a cavalcade of 400 Mahindra Maxximo mini trucks fielded by Bharatiya Janta Party in 2014 to vanquish its opponents in the impending election. The chariots were equipped with amplifiers, loudspeakers and large LCD screens that showed Narendra Modi's speeches. The Namo Ratha carried the message of Modi all over UP and enabled BJP to win 71 Lok Sabha seats in the state.

In November 2016, the UP chief minister Akhilesh Yadav also launched his Vikas Ratha with an eye on 2017 state elections. The Chariot was made of a ten-wheeled bullet proof Mercedes Benz bus with hydraulic light, reclining chairs, LCD screens, sound system, wash room, pantry, wifi, etc. The *ratha yātrā* was conducted to drum up support for his development agenda. As a had omen his chariot broke down soon after take off and he suffered a humiliating defeat.

NOTES

1. *Epigraphia Indica*, vol. 8, p. 48.
2. Samuel Beal, *Buddhist Records of the Western World*, pp. 82-3.
3. गौ-हस्ति-रथ-अश्व पत्ति-सम्पन्न स्कन्धावार-रात्। *Epigraphia Indica*, vol. 5, p. 210.

4. *Indian Antiquary*, vol. 13, p. 140.
5. R.C. Majumdar, *History of Bengal*, vol. I, p. 279.
6. चतुरंगचमूप्रचारम्। *Corpus Inscriptionum Indicarum*, vol. 4, p. 293.
7. A.L. Basham, *The Wonder that was India*, p. 131.
8. P.C. Chakravarty, *The Art of War in Ancient India*, p. 26.
9. Ibid., p. 26.
10. A. Macdonell, *A History of Sanskrit Literature*, p. 168.
11. *Mānasāra*, 36.43.
12. Ibid., 32.35.
13. Ibid., 9.192.
14. Ibid., 3.9.
15. Ibid., 37.86-89.
16. हस्तिरथेन सह प्रदत्तः। *Corpus Inscriptionum Indicarum*, vol. 4, p. 621-2.

Bibliography

ORIGINAL SOURCES

Abhidhāna Cintāmaṇi (Hemcandra), ed. Nemichandra Shastri, Chaukhamba, Varanasi, 1964.

Abhijñānaśākuntalam, (Kālidāsa), ed. Nemichandra Shastri, Chaukhamba, Varanasi, 1964.

Agni Purāṇa , tr., J.L. Sastri, Motilal Banarsidass, Delhi, 1984.

——, ed. and tr. T. Jha and G. Tripathi, Hindi Sahitya Sammelan, Allahabad, 1998.

Aitareya Brāhmaṇa, English tr. M. Haug, Bharatiya Publishing Home, Delhi, 1976.

Aṅguttara Nikāya, ed. and tr. Swami Dwarkadas Shastri, Bauddha Bharati, Varanasi, 2002.

Amarakośa, ed. Hargovind Sastri, Chowkhamba Sanskrit Series, Varanasi, 1970.

Angutara Nikāya, ed. and tr. Swami Dwarkadas Sastri, Baudha Bharati, Varanasi, 2002.

Aparājitapṛchhā, ed. P.A. Mankad, GOS No. 114, Baroda, 1950.

Āpastamba Śulbasūtra, ed. Satyaprakash and R.P. Sharma, The Research Institute of Ancient Scientific Studies, New Delhi, 1968.

Arthaśāstra, ed. and tr. R.P. Kangle, University of Bombay, Bombay, 1965-72.

Aṣṭādhyāyī, ed. and tr. S.C. Basu, Motilal Banarsidass, Delhi, 1962.

Āśvalāyana Grihyasūtra, tr. and ed. F. Max Müller, Oxford, 1886.

Aśvaśāstram, ed. T. Chandrasekharan, Govt. of Madras, 1952.

Atharvaveda Saṁhitā, ed. and tr. S.D. Satvalekar, Swadhyaya Mandal, Paradi, 1985.

——, English tr. W.T. Whitney, Motilal Banarsidass, Delhi, 1962.

Baudhāyana Dharmaśutra, ed. N.K. Acharya, Vidyānidhi Prakashan, Delhi 1999.

Baudhāyana Śrautasūtra, ed. and tr. C.S. Kashikar, Motilal Banarsidass, Delhi, 2003.

Baudhāyana Śulbasūtra ed. Satyaprakash and R.P. Sharma, The Research Institute of Ancient Scientific Studies, New Delhi, 1968.

Bhāradvāja Śrautasūtra, ed and tr. C.G. Kashikar, Vaidika Samsodhana Mandala, Poona, 1964.

Bhaviṣya Purāṇa, (Vainkatesvara), Naga Publishers, Delhi, 1985.

Brahma Purāṇa, tr. and ed. J.L. Sastri, Motilal Banarsidass, Delhi, 1985.

Brahmāṇḍa Purāṇa, tr. ed. J.L. Sastri, Motilal Banarsidass, Delhi, 1984.

——, ed. Jagdish Shastri, Motilal Banarsidass, Delhi, 1973.

Bṛhat Saṁhitā, ed. and tr. A. Jha, Chowkhamba Vidya Bhavan, Varanasi, 1969.

Dhanurveda, ed. & tr. R.P. Arya, Indian Foundation of Vedic Science, Rohtak, 2015.

Gautama Dharmasūtra, ed. and tr. Narendra Kumar, Vidyanidhi Prakashan, Delhi, 2007.

Harṣacarita, ed. N.S. Press, Bombay, 1834 Śaka.

——, English tr. E.B. Cowell and F.W. Thomas, Motilal Banarsidass, Delhi, 1968.

Jaiminīya Brāhmaṇa, ed. H.W. Bodewitz, E.J. Brill, Leiden, 1973.

Jātakas, English tr. E.B. Cowell, Motilal Banarsidass, Delhi, 1973.

Kūrma Purāṇa, Geeta Press, Gorakhpur, Saṁvat 2069.

Kādambarī, English tr. C.M Ridding, Oriental Book Reprint Corporation, New Delhi, 1974.

Kāmandakīya Nītisāra, English tr. M.L. Dutt, Chowkhamba Sanskrit Series, Varanasi, 1979.

——, ed. R.L. Mitra, Calcutta, 1972.

Kaṇva Śatapatha Brāhmaṇa, ed. and tr. C.R. Swaminathan, Indira Gandhi National Centre for Arts, New Delhi, 2001.

Kathāsaritasāgara, ed. K.N. Sharma, Patna, 1960.

Kaṭhopaniṣad, ed. and tr. Svami Prajnananda Sarasvati, Chaukhamba Sanskrit Samsthan, Varanasi, 2010.

Kātyāyana Śrautasūtra, Chowkhamba Sanskrit Series Office, Benares, 1939.

——, *English tr.* H.G. Ranade, Ranade Publication Series No. I, Poona, 1978.

Lāṭyāyana Śrautasūtra, ed. and tr. H.G. Ranade, Indira Gandhi National Centre for Arts, New Delhi, 1998.

Liṅga Purāṇa, ed., Vainkatesvara, Nag Publishers, Delhi, 1991.

Mahābhārata, Geeta Press, Gorakhpur, 2024 Vikrami.

——, BORI, Poona, 1933-68.

Mahāpurāṇa (Jinasenācārya), ed. Pannalal Jain, Bhartiya Jñānapīṭha, Kashi, 1951.

Majjhima Nikāya, ed. and tr. Swami Dvarkadas Sastri, Bauddha Bharati, Varanasi, 1991.

Mālavikāgnimitram, ed. and tr. Swami Dvarkadas Sastri, Bauddha Bharati, Varanasi, 1991.

Mānasāra, tr. P.K. Acharya, Oriental Book Reprint Corporation, New Delhi, 1980.

Mānasollasa, ed. G.K. Shringondekar, Oriental Institute, Baroda, 1961.

Mānava Śrauta Sūtra, tr. Jeannette M. Van Gelder, International Academy of Indian Culture, New Delhi, 1963.

Manusmṛti, ed. Rajvira Shastri, Arsh Sahitya Prachara Trust, Delhi, 1985.

——, ed. and tr. G. Buehler, Motilal Banarsidass, Varanasi, 1975.

Mārkaṇḍeya Mahāpurāṇa, ed. and tr. M.N. Dutt, Eastern Book Linkers, Delhi, 2005.

Matsya Purāṇa, ed. Shradha Shukla, Nag Publishers, Delhi, 2004.

Mayamatam, ed. T. Ganapati Sastri, Govt. Press, Trivandrum, 1919.

Naiṣadhīya Caritam, ed. N.R. Acharya, N.S. Press, Bombay, 1962.

Nītiprakāśikā (Vaiśampāyana), ed. T. Chandrasekharan, Govt. Press, Madras, 1953.

Pañcarātra, ed. N.R. Acharya, N.S. Press, Bombay, 1962.

Pañcaviṁśa Brāhmaṇa, English tr. W. Caland, Asiatic Society, Calcutta, 1931.

Pāraskara Grihya Sūtra, tr. and ed. F. Max Müller, Oxford, 1886.

——, ed. Vainkatesvara Press, Bombay, Saṁvat 1986.

Pratijñāyogandharāyaṇa (of Bhāsa), ed. and tr. C. Upadhyaya, Nag Publishers, Delhi, 2001.

Raghuvaṁśa, ed. and tr. S. Chaturvedi (Kalidasa Granthavali), Bharat Prakashan Mandir, Aligarh, 2019.

Rāmāyaṇa, Geeta Press, Gorakhpur, 2024 Vikrami.

Ṛgveda Saṁhitā, ed. and tr. S.D. Satvalekar, Swadhyaya Mandal, Paradi, 1978.

——, English tr. Ralph T.H. Griffith, Motilal Banarsidass, Delhi, 1973.

Śatapatha Brāhmaṇa, ed. and tr. G.P. Upadhyaya, The Research Institute of Ancient Scientific Studies, New Delhi, 1969.

Śiśupālavadham, ed. and tr. Hargovind Sastri, Chowkhamba Vidya Bhavan, Varanasi, 1955.

Śiva Purāṇa, (ed. Vainkatesvara), Nag Publishers, Delhi, 1986.

Śrīmaddevībhīgavata Mahāpurāṇa, Geeta Press, Gorakhpur, Sam, 2070.

Śukla Yajurveda Saṁhitā, ed. V.L. Pansikar, Nirnaya Sagar Press, Bombay, 1912.

Śukranīti, ed. J.C. Mishra, Chaukhamba Surbharati Prakashan, Varanasi, 1998.

Śukranītisāra, tr. B.K. Sarkar, Oriental Book Reprint Corporation, New Delhi, 1975.

Saṁvidhāna Brāhmaṇa, ed. B.R. Sharma, Kendriya Sanskrit Vidyapitha, Tirupati, 1964.

Samarāṅgaṇasūtradhāra, ed. T. Ganapati Sastri, Baroda, 1966.

Sāraṅgadharapaddhati, ed. Piter Piterson, Chaukhamba Sanskrit Pratisthan, Delhi, 1987.

Skanda Mahāpurāṇa (Vainkatesvara), Nag Publishers, Delhi, 1995.

Taittirīya Saṁhitā, ed. T.N. Dharmadhikari, Vaidika Samsodhana Mandal, Poona, 1981.

Urubhaṅga, ed. and tr. C. Upadhyaya, Naga Publishers, Delhi, 2001.

Vādhūla Śrautasūtra, ed. Braj Bihari Chaube, Katyayana Vaidik Sahitya Prakashan, Hoshiarpur, 1993.

Vāyu Purāṇa, ed. and tr. Rampratap Tripathi, Hindi Sahitya Sammelan, Prayag, 1987.

Vikramorvasīyam, ed. and tr. S. Chaturvedi, (Kalidasa Granthavali), Bharat Prakashan Mandir, Aligarh, 2019.

Viṣṇu Mahāpuraṇa, ed. Shradha Shukla, Nag Publishers, Delhi, 1998.

Viṣṇu Smṛti, English tr. Julius Jolly, Motilal Banarsidass, Delhi, 1965.

Viṣṇudharmottara Purāṇa, (Vainkatesvara edn.), Nag Publishers, 2009.

——, tr. Priyabala Shah, Parimal Publications, Delhi, 2005.

Viśvakarmīya Rathalakṣaṇam, R.P. Kulkarni, Kanishka Publishars, Delhi, 1994.

Yājñavalkyasmṛti, ed. U. Pandey, Chowkhamba, Varanasi, 1967.

Yajurveda Saṁhitā, ed. and tr. S.D. Satvalekar, Swadhyaya Mandal, Paradi, 1985.

Yuktikalpataru, ed. B.L. Rajpurohit, Pratika Prakashan, Delhi, 2008.

SOME OTHER WORKS

Percy Gardner, *The Coins of Greek and Scythian Kings of Bactria and India in the British Museum*, Sagar Publications, New Delhi, 1971.

Homer, *The Iliad* (tr. Robert Fagles), Viking Penguin, New York, 1990.

——, *The Odyssey* (tr. Robert Fagles), Penguin Books India Pvt. Ltd., New Delhi, 1997

Legge, J., *The Travels of Fa-hien*, Oriental Publishers, Delhi, 1971.

Marshall, J., *Taxilā*, University Press, Cambridge, London, 1951.

——, *Mohenjo-daro and the Indus Civilization*, Indological Book House, Delhi, 1973.

M'crindle, J.W., *Ancient India as Described by Megasthenese and Arrian*, Today and Tomorrow's Printers and Publishers, New Delhi, 1972.

——, *The Invasion of India by Alexander the Great*, Today and Tomorrow's Printers and Publishers, New Delhi, 1974.

Sachau, E., *Alberuni's India*, Indialog Publications Pvt. Ltd., Delhi, 2003.

Sun Tzu, *The Art of War*, (Denma Trans. Group), Shambhola Publications, Boston, USA.

Watters, T., *On Yuan Chwangs Travels in India*, Munshiram Manoharlal, New Delhi, 1973.

SECONDARY SOURCES

Acharya, P.K., *Hindu Architecture in India and Abroad*, J.K. Publishing House, Bhopal, 1979.

Agarwal, D.P., *The Indus Civilization: An Inter-disciplinary Perspective*, Aryan Book International, New Delhi, 2007.

Agnihotri, Prabhudayal, *Patañjalikālīna Bhārata*, Bihar Rashtrabhasa Parishad, Patna, 2001.

Agrawala, V.S., *Harṣacarita Eka Sāṁskṛtika Adhyayana*, Bihar Rashtrabhasha Parishad, Patna, 1953.

——, *India as Known to Pāṇini*, University of Lucknow, Lucknow, 1953.

Bachhofer, L., *Early Indian Sculpture*, Paris, 1929.

Barua, B.M., *Bhārhut*, Indological Book Corporation, Patna, 1979.

——, *Buddha-Gayā*, Bharatiya Pulishing House, Varanasi, 1975.

Barua, D.K., *An Analytical Study of Four Nikāyas*, Rabindra Bharati University, Calcutta, 1971.

Basham, A.L., *The Wonder that was India*, Picador, London, 2004.

Basu, J., *India of the Age of Brāhmaṇas*, Sanskrit Pustak Bhandar, Calcutta, 1969.

Beal, Samuel, *Si-yu-ki Buddhist Records of the Western World* (tr.), Munshiram Manoharlal, New Delhi, 1983.

Bertman, Stephen, *Handbook to Life in Ancient Mesopotamia*, Facts on File Inc., New York, 2003.

Bradford, Alfred, S., *With Arrow, Sword and Spear: A History of Warfare in the Ancient World*, Prager Publishers, Connecticut, USA.

Chakravarty, P.C., *Art of War in Ancient India*, University of Dacca, Dacca, 1941.

Chakravorti, Ranabir, *Warfare for Wealth*, Firma KLM Pvt. Ltd., Calcutta, 1986.

Chattopadhyaya, D., *History of Science and Technology in Ancient India: The Beginnings*, Firma KLM Pvt. Ltd., Calcutta, 1986.

Choodamani Nandagopal & Vatsala Iyengar, *Temple Treasures*, vol. III, The Craft Council of Karnataka, Bangalore, 2002.

Cotterell, Arthur, *Chariot*, Pimlico, London, 2004.

Cummins, Joseph, *From Chariot to Flintlocks*, Fair Wind Press, Massachusetts, 2008.

Cunningham, A., *The Stūpa of Bhārhut*, Indological Book House, Varanasi, 1962.

——, *The Bhilsā Topes*, Indological Book House, Varanasi, 1966.

Das, A.C., *Ṛgvedic Culture*, Bharatiya Publishing House, Delhi, 1979.

Date, G.T., *The Art of War in Ancient India*, Calcutta, 1943.

David, R., *Handbook of Life in Ancient Egypt*, Facts on File Inc., New York, 1998.

David W. Del Testa, *Global History*, vol. 1, Sharpe Reference, New York, 2004.

Deekshit, V.R.R., *War in Ancient India*, Motilal Banarsidass, Delhi, 1987.

Dey, S.N., *Indian Life in Śukla Yajurveda*, Firma KLM, Calcutta, 1985.

Dhavalikar, M.K., *Ajantā: A Cultural Study*, Deccan College, Poona, 1965.

——, *Indian Protohistory*, Books and Books, New Delhi, 1997.

Dutt, R.C., *A History of Civilization in Ancient India*, Cosmo Publications, New Delhi, 2000.

Erman, Adolph, *Life in Ancient Egypt*, Dover Publications, Inc., New York, 1971.

Fergusson, J., *Tree and Serpent Worship*, Indological Book House, Delhi, 1971.

Fields, Nic, *Bronze Age War Chariots*, Osprey Publishing Ltd., Oxford, 2006.

Frawley, David, *Gods, Sages and Kings-Vedic Secrets of Ancient Civilization*, Motilal Banarsidass, Delhi, 1999.

——, *The Myth of the Aryan Invasion of India*, Voice of India, Delhi, 1994.

Futrell, A., *The Roman Games*, Blockwell Publishing, Malden, USA.

Gode, P.K., *Studies in Indian Cultural History*, vol. 1, Vishveshvaranand Vedic Research Institute, Hoshiarpur, 1961; vol. 2, P.K. Gode Collected Works Publications Committee, Poona, 1960.

Gordon Childe, V., *New Light on the Most Ancient East*, Routledge and Kegan Paul Ltd., London, 1958.

Gordon, D.H., *The Pre-historic Background of Indian Culture*, Bhulabhai Memorial Institute, Bombay, 1954.

Griffiths, J., *The Paintings in the Buddhist Cave Temples of Ajantā*, Secretary of State for India in Council, London, 1896.

Gupta, D.K., *Society and Culture in the Time of Daṇḍin*, Meharchand Lachhmandass, Delhi, 1972.

Habib, Irfan, *The Indus Civilization*, Tulika Books, New Delhi, 2002.

Hopkins, E.W., *The Social and Military Position of the Ruling Caste in Ancient India*, Bharat Bharati, Varanasi, 1972.

Huntington, S.L., *The Art of Ancient India*, Weather Hill, New York, 1985.

Ingholt, H., *Gandhāran Art in Pakistan*, Lyons Islay, New York, 1957.

Joshi, N.P., *Life in Ancient Uttarāpatha*, Benares, 1967.

Kaegi, Adolph, *Life in Ancient India*, Calcutta, 1950.

Kak, R.C., *Ancient Monuments of Kashmir*, The Indian Society, London, 1933.

Kanaksabhai, V., *The Tamils Eighteen Hundred Years Ago*, Asian Educational Services, New Delhi, 1979.

Kane, P.V., *History of Dharmaśāstra* (5 vols.), BORI, Poona, 1941.

Keegan, John, *A History of Warfare*, Vintage Books, New York, 1994.

Kosambi, D.D., *The Culture and Civilization of Ancient India in Historical Outline*, New Delhi, 1989.

Kriwaczek, Paul, *Mesopotamia and the Birth of Civilization*, Thomas Dunne Books, New York, 2012.

Kulkarni, R.P., *Viśvakarmīya Rathalakṣaṇam*, Kaniska Publishers and Distrubutors, Delhi, 1994.

Lal, B.B., *The Earliest Civilization of South Asia*, Aryan Books International, New Delhi, 1997.

——, *The Homeland of the Āryans*, Aryan Books International, New Delhi, 2005.

Longhurst, A., *The Buddhist Antiquities of Nāgārjunikoṇḍa*, Madras Presidency, MASI, No. 54, 1938.

Macdonell, A.A., *A History of Sanskrit Literature*, London, 1913.

Macdonell, A.A. and A.B. Keith, *Vedic Index of Names and Subjects*, Motilal Banarsidass, Varanasi, 1958.

Mackay, E.J.H., *Further Excavations at Mohenjo-daro*, Indological Book Corporation, New Delhi, 1976.

——, *Early Indus Civilization*, Indological Book Corporation, Delhi, 1976.

Macqueen, J.G., *The Hittites*, Thames and Hudson Ltd., London, 1986.

Maisey, F.C., *Sāñcī and Its Remains*, Indological Book House, Delhi, 1972.

Majumdar, B.K., *The Military System in Ancient India*, Firma K.L. Mukhopadhyay, Calcutta, 1960.

Majumdar, R.C., *The History of Bengal*, N.V. Publications, Patna, 1971.

Majumdar, R.C. (ed.), *The Vedic Age*, Bharatiya Vidya Bhavan, Bombay, 1965.

——, *The Age of Imperial Unity*, Bharatiya Vidya Bhavan, Bombay, 1960.

——, *The Classical Age*, Bharatiya Vidya Bhavan, Bombay, 1962.

Marshall, J., *A Guide to Sāñcī*, Manager of Publication, Delhi, 1936.

——, *Mohenjo-daro and the Indus Civilization*, Indological Book House, Delhi, 1973.

Marshall, J. and Foucher A., *The Monuments of Sāñcī*, Archaeological Survey of India, Calcutta.

Mertz, Barbara, *Temples Tombs and Hieroglyphs: A Popular History of Ancient Egypt*, Harper Collins Publishers, New York, 2007.

Mireaux, E., *Daily Life in the Time of Homer*, tr. George Allen and Unwin Ltd., London, 1959.

Mitra, R.L., *The Antiquities of Orissa*, vol. 1, Calcutta, 1895.

——, Indo-Aryans, Cosmo Publications, Delhi, 1999.

Oppert, Gustava, *On the Weapons, Army Organization and Political Maxims of the Ancient Hindus*, New Order Books, Ahmedabad, 1967.

Pandey, G.C., *Dimensions of Ancient Indian Social History*, vol. II, Books and Books, Delhi, 1985.

Pant, G.N., *Studies in Indian Weapons and Warfare*, Army Educational Store, New Delhi, 1970.

——, *Indian Archery*, Agam Kala Prakashan, Delhi, 1978.

Piggott, S., *Prehistoric India.*, Cassell and Company Ltd., London, 1962.

——, *Wagon, Chariot and Carriage*, Thames and Hudson, London, 1992.

Postgate, J.N., *Early Mesopotamia*, Routledge, London, 1992.

Ragozin, Zanaide, A., *History of Vedic India*, Mittal Publications, Dlehi, 1984.

Ram Gopal, *India in Vedic Kalpasūtras*, Delhi, 1959.

Rangarajan, L.N., *The Arthaśāstra*, Penguin Book India, New Delhi, 1992.

Rapson, E.J., *Cambridge History of India*, vol. 1, S. Chand and Co., Delhi, 1955.

Raychaudhuri, H.C., *Political History of Ancient India*, University of Calcutta, Calcutta, 1953.

Reeves, Nicholas, *The Complete Tutankhamun*, Thames and Hudson Ltd., London, 1990.

Robinson, H. Russel, *Oriental Armour*, London, 1967.

Roy, B.P., *Political Ideas and Institutions in the Mahābhārata*, Punthi Pustaka, Calcutta, 1975.

Schlingloff, Dieter, *Studies in the Ajanta Paintings*, Ajanta Publications (India), Delhi, 1988.

Sharma, R.C., *The Splendour of Mathurā Art and Museum*, R.K. Print World, New Delhi, 1994.

Sharma, R.N., *Culture and Civilization as Revealed in Śrautasūtras*, Nag Publishers, Delhi, 1977.

Sharma R.S., *New Light on Ancient Indian Society and Economy*, Manaktalas, Bombay, 1966.

——, *Aspects of Political Ideas and Institutions in Ancient India*, Motilal Banarsidass, Delhi, 1999.

Sharma, Subhra, *Life in the Upanisads*, Abhinava Publications, New Delhi, 1985.

Shastri, A.M., *India as Seen in the Bṛhatsaṁhitā of Varāhamihira*, Motilal Banarsidass, Delhi, 1969.

Shastri, K.A. Nilakanta, *History of South India*, Oxford University Press, London, 1966.

Shaw, Garry J., *The Pharaoh: The Life at Court and Campaign*, Thames and Hudson Ltd., London, 2012.

Simon, Anglim, *Fighting Techniques of the Ancient World*, Thomas Dunne Books, New York, 2002.

Singh, Sabhapati, *Prācīna Bhārata mein Sainya Vyavasthā*, Durga Publications, Delhi, 1990.

Singh, S.D., *Ancient Indian Warfare with Special Reference to Vedic Period*, E.J. Brill, Lieden, 1965.

Smith, V.A., *Early History of India*, Oxford, 1914.

——, *The Jain Stūpa and other Antiquities of Mathurā*, Indological Book Home, Delhi, 1969

Sivaramamurti, C., *Amarāvatī*, Govt. Press, Madras, 1956.

Sparreboom, M., *Chariots in the Veda*, Leiden, 1985.

Srivastava, A.L., *Life in Sāñcī Sculptures*, Abhinava Publications, New Delhi, 1983.

Srivastava, Priya, *Prācīna Baudha Granthoṁ mein Varṇita Dhātu or Dhātukarma*.

Staal, Frits, *Discovering the Vedas*, Penguin Books, New Delhi, 2008.

Subrahmanian, N., *Saṅgam Polity*, Asia Publishing House, New Delhi, 1966.

Sulimirski, T., *The Sarmatians*, Thames and Hudson, London, 1970.

Thapar, Romila, *Ancient Indian Social History*, Orient Longman Limited, New Delhi, 1978.

——, *The Penguin History of Early India*, Penguin Books, London, 2002.

Thapliyal, U.P., *The Dhvaja*, B.R. Publishing Corporation, Delhi, 1983.

——, *Historical Perspectives of Warfare in india: Some Morale and Material Determinants* (ed. S.N. Prasad), Centre for Studies in Civilization, New Delhi, 2002

——, *Warfare in Ancient India: Organizational and Operational Dimensions*, Manohar, New Delhi, 2010.

Tripathi, R.S., *History of Ancient India*, Motilal Banarsidass, Delhi, 1981.

Vādhūla Śrautasūtra, ed. Braj Bihari Chaube, Katyayana Vaidik Sahitya Prakashan, Hoshiarpur, 199 3.

Wells, H.G., *The Outline of History*, vol. I, Low Price Publications, Delhi, 1998.

Will Durant, *The Life of Greece*, Simon and Schuster, New York, 1966.

Yadin, Y., *The Art of Warfare in Biblical Lands*, McGraw Hill, New York, 1963.

Yashpal, *A Cultural Study of Early Pāli Tripitakas* (2 vols.), Kalinga Publications, Delhi, 1999.

Yazdani, G., *Ajantā*, Oxford University Press, London, 1930.

Index